Jin Wang

Pattern Effect Mitigation Techniques for All-Optical Wavelength Converters Based on Semiconductor Optical Amplifiers

Karlsruhe Series in Photonics & Communications, Vol. 3
Edited by Prof. J. Leuthold and Prof. W. Freude

Universität Karlsruhe (TH), Institute of High-Frequency and Quantum Electronics (IHQ), Germany

Pattern Effect Mitigation Techniques for All-Optical Wavelength Converters Based on Semiconductor Optical Amplifiers

by
Jin Wang

universitätsverlag karlsruhe

Dissertation, Universität Karlsruhe (TH)
Fakultät für Elektrotechnik und Informationstechnik, 2008

Impressum

Universitätsverlag Karlsruhe
c/o Universitätsbibliothek
Straße am Forum 2
D-76131 Karlsruhe
www.uvka.de

Dieses Werk ist unter folgender Creative Commons-Lizenz
lizenziert: http://creativecommons.org/licenses/by-nc-nd/2.0/de/

Universitätsverlag Karlsruhe 2008
Print on Demand

ISSN: 1865-1100
ISBN: 978-3-86644-276-4

Pattern Effect Mitigation Techniques for All-Optical Wavelength Converters Based on Semiconductor Optical Amplifiers

Zur Erlangung des akademischen Grades eines

DOKTOR-INGENIEURS

von der Fakultät für
Elektrotechnik und Informationstechnik
der Universität Karlsruhe (TH)
genehmigte

DISSERTATION

von

Dip.-Ing. Jin Wang
geb. in Jiangsu, China

Tag der mündlichen Prüfung   09.06.2008
Hauptreferent   Prof. Dr.-Ing. Dr. h. c. W. Freude
Korreferent   Prof. Dr. sc. nat. J. Leuthold

To my family…

Zusammenfassung-Abstract

In den letzten Jahren hat sich die Nachfrage nach mehr Bandbreite in Telekommunikationsnetzwerken erheblich erhöht. Dieser Trend wird sich erwartungsgemäß auch in den nächsten Jahren nicht ändern, da neue Dienstleistungen im Multimedia-Bereich immer mehr Bandbreite benötigen werden. Um eine zunehmende Bandbreite der optischen Faser auszunutzen, werden sowohl Wellenlängenmultiplex- (WDM, wavelength division multiplexing) als auch Zeitmultiplex-Verfahren (TDM, time division multiplexing) angewandt. Bei beiden grundlegenden Techniken spielen Wellenlängenkonverter (wavelength converters), die eintreffende optische Signale einer Wellenlänge auf eine zweite aufprägen, eine wichtige Rolle. Ein Wellenlängenkonverter muss insbesondere effizient in Hinblick auf fehlerfrei kovertierte Signale bei möglichst geringen optischen Eingangsleistungen sein. Wellenlängenkonverter sollen zudem kompakt, einfach aufgebaut und damit kostengünstig herstellbar sein.

Für Wellenlängen-Konversion werden heutzutage optisch-elektrisch-optische (O-E-O) Wandlungen durchgeführt. Die Konverter sind voluminös und weisen eine hohe Leistungsaufnahme auf. Der Hauptvorteil eines elektrischen Anteils in der Wandlung ist die gleichzeitig erzielbare Signal "3R"-Regeneration: Re-amplification, Re-shaping, Re-timing (Verstärkung, Wiederherstellung einer Formtreue, und Resynchronisierung). Dabei werden gängige und bewährte elektronische Schaltkreise eingesetzt. Trotz grosser Fortschritte der letzten Jahre ist komplexe Elektronik mit >100 Mbit/s Datendurchsätzen sehr teuer.

Von voll-optischen (all-optical) Wellenlängenkonvertern verspricht man sich durch eine Konversion direkt im Optischen, und Vermeiden einer elektrischen Konversion kostengünstigere, kompaktere und bei Datenraten mit >100 Mbit/s funktionierende Komponenten. Damit können Einschränkungen durch die Blockierung von Wellenlängenkanälen ("blocking") in transparenten Netzwerken im Optischen überwunden werden, bei Wieder-Allozierbarkeit ursprünglicher Wellenlängen. Flexibles Routing und Schalten zwischen globalen und lokalen Netzwerken werden durchführbar durch verschiedene Konzepte, die optische Schalter, optische Burst- und Paket-Vermittlung ("optical labeling") sowie phasensensitive Formate mit Interferometern nutzen.

Eine für voll-optische Wandlung zentrale Komponente ist der optische Halbleiterverstärker SOA (semiconductor optical amplifier). SOA-basierte Wellenlängenkonverter sind kompakt, haben niedrige Leistungsaufnahmen, und können bei höchsten Übertragungsgeschwindigkeiten verwendet werden. Wellenlängenkonversion im Optischen mit 320 Gbit/s wurden bereits demonstriert. SOAs enthalten in ihrer Funktionsweise auch elektronische Anteile: Jedoch laufen Transport- und Rekombinationsprozesse bei der genutzten stimulierten Emission von Photonen sehr schnell ab (typisch <100 ps), unter anderem durch Beschränkung auf kleine aktive Volumina. Diese sind bestimmt durch die sehr kleinen Abmessungen von wellenleitenden Kernen und ihrer direkten Umgebungen (einige 100 nm bis 1 µm).

Die SOA-basierte Wellenlängenkonversion basiert zentral auf einer Veränderung des lokalen Verstärkungsfaktors ("gain") für ein optisches Signal in der aktiven Region in Verbindung mit einer Modulation des Brechungsindexes bei Änderung der lokalen Konzentration freier Ladungsträger. Mehrere optische Signale können gleichzeitig durch den SOA propagieren mit der Möglichkeit einer gezielten gegenseitigen Beeinflussung und Wellenlängenkonversion. Im SOA finden eine Vielzahl dynamischer Prozesse während der Modulation statt. Von Interesse sind insbesondere Zeitkonstanten für Inter- und Intraband-Ladungsträgerprozesse bezüglich genutzter elektronischer Valenz- und Leitungszustände. Weiter führen verschiedene konkrete Abfolgen von Signal-Bit-Mustern ("bit patterns") zu verschiedenen, jedoch deterministischen gewandelten Signalverläufen ("pattern effect").

In dieser Dissertation werden Modelle zu SOA-basierten, voll-optischen Wellenlängenkonvertern vorgestellt. Mit Hilfe von Simulationen werden generische SOAs charakterisiert wie auch Optimierungen von SOAs für spezialisierte Anwendungen durchgeführt. Besonderes Augenmerk gelten signal-verschlechternden "pattern effects", die es zu verstehen und minimieren gilt. Simulations-theoretische Modelle werden in Einklang gebracht mit experimentellen Ergebnissen an ausgewählten realen Bauteilen ("devices") unter Laborbedingungen in einem "optical test loop".

Die Dissertation ist wie folgt strukturiert: In Kapitel 1 wird eine Einleitung zu optisch verstärkenden und wellenlängenschiebenden Halbleiterverstärkern SOAs gegeben.

In Kapitel 2 stellt numerische Modelle und Verfahren zur Pulspropagation in SOAs vor, mit Fokus auf Gewinnsättigung ("gain saturation") und Zeitkonstanten für Intra- und Interband-Ladungsdynamiken

Kapitel 3 führt Experimente zur Kreuzgewinnmodulation XGM (cross-gain-modulation) und Kreuzphasenmodulation XPM (cross-phase-modulation) und deren Erklärung mit Simulationsmodellen vor.

Kapitel 4-6 geben einen Überblick über unterschiedliche Schemata und Aufbauten zur Realisierung von Wellenlängenkonversion. In Kapitel 4 wird eine Verminderung des optischen Pattern-Effects mit Hilfe eines speziellen optischen Filteraufbaus PROF (pulse reformatting optical filter) in einer Delay-Interferometer-Konfiguration diskutiert. In Kapitel 5 kommt ein SOA-integrierender Sagnac-Ring zum Einsatz. In Kapitel 6 wird die konventionelle direkten Modulation im OOK-Verfahren (on-off-keying) erweitert um phasen-enkodierende Elemente in DPSK-Verfahren (differential-phase-shift-keying).

Kapitel 7 beschliesst die Dissertationsarbeit mit einer Zusammenfassung und Vorschlägen für zukünftige weitere Forschungsgesichtspunkte.

Abstract

The demand for bandwidth in telecommunication network has been increasing significantly in the last few years. It is to be expected that also in the next few years multimedia services will further increase the bandwidth requirement. To utilize the entire bandwidth of the optical fibre, wavelength division multiplexing (WDM) and time-division multiplexing (TDM) techniques have been applied. In these two primary techniques, wavelength converters that translate optical signals of one wavelength into optical signals of another wavelength have become key devices. In general, a wavelength converter has to be efficient, meaning that with low signal powers an error free converted signal can be obtained. Also, the wavelength converters have to be small, compact, and as simple (cheap) as possible.

Presently, large footprint optical-electrical-optical (o-e-o) translator units with large power consumption are used to perform wavelength conversion in optical cross-connects. Advantages of o-e-o methods are their inherent 3R (re-amplification, re-shaping, re-timing) regenerative capabilities and maturity. Conversely, the promise of all-optical wavelength conversion is the scalability to very high bit rates. All-optical wavelength converters can overcome wavelength blocking issues in next generation transparent networks and make possible reuse of the local wavelengths. All-optical wavelength converters can also enable flexible routing and switching in the global and local networks, e.g. optical circuit-switching, optical burst-switching and optical packet-switching.

Semiconductor optical amplifiers (SOA) are closest to practical realization of all-optical wavelength converters. SOA-based all-optical wavelength converters are compact, have low power consumption, and can be possibly operated at high speed. Indeed, a 320 Gbit/s SOA-based all-optical wavelength conversion has already been demonstrated. The advantages of using SOAs arise from the large number of stimulated emitted photons and free carriers, which are confined in a small active volume.

The SOA-based wavelength conversion works mostly via the variation of the gain and refractive index induced by an optical signal in the active region. The optical signal incident into the active region modifies the free carrier concentration. Thus, the optical gain and the refractive index within the active region are modulated. Other optical signals propagating simultaneously through the SOA also see these modulations of the gain and refractive index. Thus, the information is transferred to another wavelength. In SOAs, a rich variety of dynamic processes drive the operation. These processes include the carrier dynamics between conduction band and valence band (interband dynamics) as well as inside of conduction band or valence band (intraband dynamics). They both affect the gain and the refractive index of the SOA, and thus the operation of the SOA-based wavelength converter. In addition, the fact that each of these effects has a specific lifetime leads to pattern dependent effects in the processed signals. It is therefore very important to develop numerical models for the SOA to understand its behavior in wavelength conversion.

In this dissertation, an SOA model is developed to investigate high-speed SOA-based all-optical wavelength converter. With the help of the SOA model, characterization of the generic

SOA and optimization of SOA in different schemes are performed. In particular, the model has been used to investigate and introduce new SOA pattern effect mitigation techniques. The theoretical predictions are supported and verified by experiments.

The dissertation is structured as follows: In Chapter 1, an introduction to the SOA and SOA-based wavelength converters is given. Chapter 2 is dedicated to the numerical modeling of the pulse propagation, gain saturation, interband and intraband carrier dynamics inside the SOA. In Chapter 3, the SOA gain and phase responses in the cross-gain modulation experiments are shown and discussed with the help of the simulation using the SOA model. Different wavelength conversion schemes are discussed in Chapter 4-6. In Chapter 4 we discuss optical filter-assisted wavelength conversion. We first discuss a pattern effect mitigation technique by using a delay interferometer filter after an SOA. We then discuss pattern effect mitigation techniques by using a general pulse reformatting optical filter, which reformats the output signal after the SOA into a new signal with a predetermined shape. The pattern effect mitigation technique in wavelength conversion schemes using an SOA-based Sagnac loop is discussed in Chapter 5. While the former Chapters discuss wavelength conversion schemes for on-off keying (OOK) optical signals, Chapter 6 proposes the wavelength conversion scheme for differential phase-shift keying (DPSK) modulated signal. A summary of the work and an outlook on future research are given in Chapter 7.

Contents

List of Symbols and Acronyms

Constants

c	speed of light in vacuum	2.99792458×10^{8} m/s
$\hbar$	reduced Planck constant	$1.0545727 \times 10^{-34}$ Js
h	Planck constant	$6.6260755 \times 10^{-34}$ Js
π	pi	3.141592
q	elementary charge	$1.60217733 \times 10^{-19}$ C
ε_0	permittivity of free space	$8.854187818 \times 10^{-12}\ \dfrac{\text{As}}{\text{Vm}}$
μ_0	permeability of free space	$4\pi \times 10^{-7}\ \dfrac{\text{Vs}}{\text{Am}}$

Acronyms

ASE	amplified spontaneous emission
BER	bit-error probability (usually called bit-error rate)
BERT	bit-error rate test
BSOF	blue-shifted optical filter
CB	conduction band
CH	carrier heating
CW	continuous wave
CDR	carrier density response
DPSK	differential phase-shift keying
DI	delay interferometer
EAM	electro-absorption modulator
EDFA	erbium doped fiber amplifier
FBG	fibre Bragg grating
FWHM	full width at half maximum
FWM	four-wave mixing

GVD	group velocity dispersion
HNLF	highly nonlinear fibre
MZI	Mach-Zehnder interferometer
NRZ	non-return-to-zero
OD	optical delay
OOK	on-off keying
OSNR	optical signal-to-noise ratio
PRBS	pseudo random bit sequence
PROF	pulse reformatting optical filter
RSOF	red-shifted optical filter
RZ	return-to-zero
SOA	semiconductor optical amplifier
SHB	spectral hole burning
SNR	signal-to-noise ratio
SPM	self-phase modulation
TPA	two-photon absorption
VB	valence band
VOA	variable optical attenuator
XGM	cross-gain modulation
XPM	cross-phase modulation

1 Introduction

1.1 All-Optical Wavelength Conversion with Semiconductor Optical Amplifiers

A wavelength converter is a device which transforms information on an optical carrier at one wavelength onto another optical carrier at a new wavelength. It is a critical function in future optical network. It may overcome wavelength contention issues[1] in next generation transparent networks [90]. In general a wavelength converter has to be efficient, meaning that with low signal powers an error free converted signal can be obtained. From a usual operator point of view, the wavelength converters have to be small, compact, and as simple (cheap) as possible.

In future high speed networks, signal processing functionalities will have to be solved all-optically to overcome the electronics bottleneck. Some all-optical approaches utilizing nonlinear interactions in highly nonlinear media have been proposed. The nonlinear medium can be for example a semiconductor optical amplifier (SOA) [10], [12], [30], [43], [71], [75], or a piece of highly nonlinear fibre (HNLF) [67], [83], or a nonlinear waveguide using other materials [1], [34], [74]. Among them, SOA devices provide stronger nonlinear interactions therefore allowing for short device lengths.

In the following, we will discuss the fundamentals of all-optical wavelength conversion using SOAs. The basic schemes using SOAs will be introduced.

1.1.1 What is a Semiconductor Optical Amplifier?

Semiconductor optical amplifiers (SOAs) are amplifiers which use a semiconductor as the gain medium. These amplifiers have a similar structure to Fabry-Perot laser diodes but with anti-reflection elements at the end faces. SOAs are typically made from group III-V compound direct bandgap semiconductors such as GaAs/AlGaAs, InP/InGaAs, InP/InGaAsP and InP/InAlGaAs. Such amplifiers are often used in telecommunication systems in the form of fiber-pigtailed components, operating at signal wavelengths between 0.85 μm and 1.6 μm. SOAs are potentially less expensive than erbium doped fiber amplifier (EDFA) and can be integrated with semiconductor lasers, modulators, etc. However, the drawbacks of SOAs are challenging polarization dependences and a higher noise figure. Practically, the polarization dependence in the SOA can be reduced by an optimum structural design [68].

A schematic diagram of a heterostructure SOA is given in Fig. 1.1. The active region, imparting gain to the input signal, is buried between the p- and n-doped layers, while the length of the active region is L. An external electrical current injects charge carriers into the active

---

[1] A wavelength contention problem can arise when two different channels of data from two input fibers are carried by input optical lights of identical or similar wavelengths, and the application requires sending both channels out in a single output fiber by wavelength division multiplexing.

region and provides a gain to the optical input signal. An SOA typically has an amplifier gain of up to 30 dB.

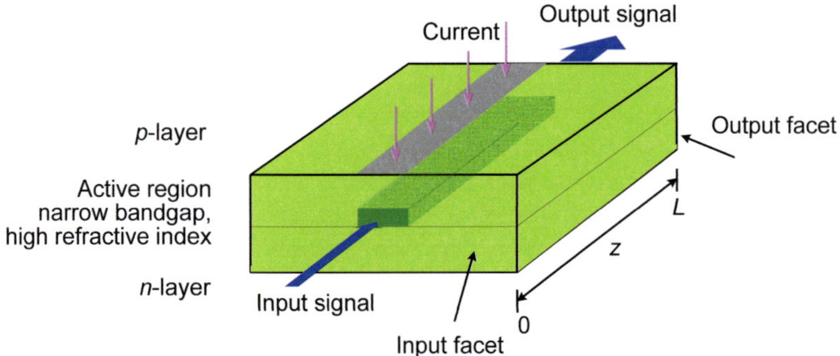

Fig. 1.1. Schematic diagram of an SOA.

Semiconductor amplifiers interact with the light, i.e. photons, in terms of electronic transitions. The transition between a high energy level W_2 in the conduction band (CB) and a lower energy level W_1 in the valence band (VB) can be radiative by emission of a photon with energy $hf = W_2 - W_1$ (h: Planck's constant, f: frequency), or non-radiative (such as thermal vibrations of the crystal lattice, Auger recombination). Three types of transitions, namely stimulated absorption, spontaneous emission and stimulated emission, are the basic mechanisms for all lasers and optical amplifiers. They are.

- **Stimulated absorption:** If an electromagnetic field exists, a photon with energy $hf = W_2 - W_1$ can be absorbed. Meanwhile, an electron can make an upwards transition from the lower energy level W_1 to the higher energy level W_2. The stimulated absorption rate depends on the electromagnetic energy density, and on the number of the electrons in the CB and the number of holes in the VB.

- **Spontaneous emission:** An electron in the CB can with a certain probability undergo a transition to the lower energy level W_1 spontaneously, while emitting a photon with energy $hf = W_2 - W_1$, or loosing the transition energy through phonons or collisions. Obviously, the probability of the spontaneous emission is dependent on the number of the electron and hole pairs. These "spontaneously" emitted photons will be found with equal probability in any possible modes of the electromagnetic field. Thus a spontaneously emitted photon is regarded as a noise signal, since it represents a field with a random phase and a random direction.

- **Stimulated emission:** An incident photon can also induce with a certain probability a transition of the electron from the high energy level W_2 to the low energy level W_1. In this process a second photon is created. In contrast to spontaneous emission processes, this second photon is identical in all respects to the inducing photon (identical phase, frequency and direction). As does the stimulated absorption, the stimulated emission rate also depends on the incident electromagnetic energy density.

The SOA shows high nonlinearity with fast transient time. However, the high nonlinearity in the SOA is also accompanied with phase changes, which can distort the signals. This nonlinearity presents a most severe problem for optical communication applications, if the SOAs are used as "linear" optical amplifiers. Yet, high optical nonlinearity makes SOAs attractive for all-optical signal processing like all-optical switching and wavelength conversion. Indeed, the nonlinear gain and phase effects in an SOA, usually termed as cross-gain modulation (XGM) and cross-phase modulation (XPM), are widely used to perform wavelength conversion.

The XGM and XPM effects in an SOA are schematically shown in Fig. 1.2.

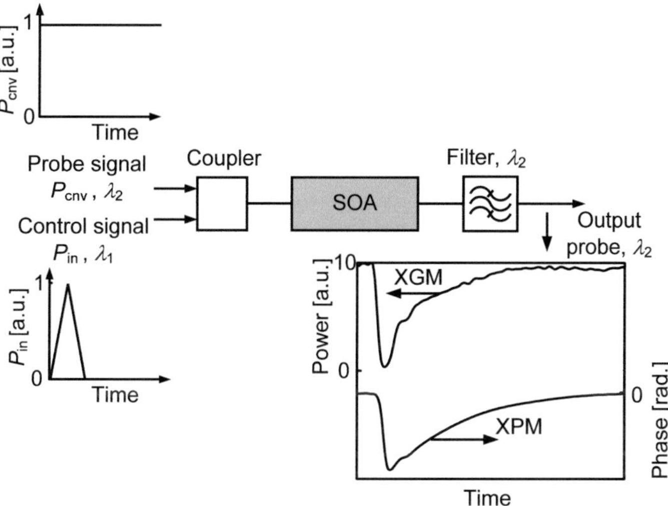

Fig. 1.2. XGM and XPM effects in an SOA. As an optical pulse of the input control signal at λ_2 is launched in the SOA, XGM and XPM effects are visualized by a cw probe signal at λ_1 at the output of the SOA, see left and right y-axes of the output signal.

- **XGM**: As a strong control signal is launched into an SOA, it depletes the carriers inside the SOA and saturates the gain of the SOA. If a weak probe signal (usually at a new wavelength) is also present in the SOA, the amplification of the probe signal is affected by the strong control signal, see the output power in Fig. 1.2. Thereby, the probe signal is modulated by the control signal. This effect is called cross-gain modulation (XGM).

- **XPM**: As the control signal causes a change of the carrier density inside the SOA, it also induces a change of the refractive index and a phase shift of the probe signal when passing through the SOA, see the output phase in Fig. 1.2. This effect is called cross-phase modulation (XPM). The XPM effect is usually measured in an interferometric configuration.

In an SOA, the XGM and XPM are related phenomenologically by a linewidth enhancement factor[2] α_H, also called Henry factor [27]. The α_H-factor is defined as the ratio between the changes of the gain and the refractive index. In most publications [1], [27] and [57], the α_H-factor is assumed to be constant. This simplification indeed is valid only over a limited spectral range and for sufficiently small carrier density modulations in the SOA.

After the control signal has died out, carriers recover. The carrier recovery process includes inter- and intraband carrier dynamics[3]. The interband carrier dynamics limits the operation speed of the SOA. Especially, when subsequent control pulses are launched into a slow SOA (with a large carrier recovery time), the carrier density is progressively reduced. It recovers back to different levels and the amplifier gain also varies for different pulses, depending on the bit pattern seen by the SOA. This effect is usually called "pattern effect".

In this work, we will analyze the various schemes using SOAs to perform all-optical wavelength conversion. In particular, pattern effect mitigation techniques are investigated and implemented to operate SOA-based wavelength converters successfully (with bit-error probability (BER) down to 10^{-9}) at a speed above 40 Gbit/s.

1.1.2 All-Optical Wavelength Conversion Schemes Using SOAs

All-optical wavelength conversion with SOAs in interferometric configurations [12], [38], [43], [44], [47], [50], [75], [78] have been successfully demonstrated, e.g. at a speed up to 320 Gbit/s [50]. In addition to demonstrations of high speed, reference [38] showed an ideal extinction ratio, reference [44] showed 2R (re-amplification, re-shaping) property over a 16,800 km transmission and reference [43] showed 3R (re-amplification, re-shaping and re-timing) regenerative capabilities. We now discuss four generic configurations, namely an SOA-based Mach-Zehnder interferometer (MZI), an SOA-based Sagnac interferometer, an SOA followed by a delay interferometer (DI), and an SOA-based wavelength converter assisted by an optical filter.

SOAs in Mach-Zehnder Interferometer

An all-optical wavelength converter using a symmetric MZI is depicted in Fig. 1.3. The MZI comprises two equally biased SOAs (SOA1 and SOA2), which ideally provide identical gains. The respective coupler introduces the control signal into the SOA section on the MZI arms. In the switched-off state, when the control signal is absent, the probe signal is directed towards the sum output port P_Σ, supposing that the symmetric phase relations are correctly adjusted. In the switched-on state, a control signal P_{in} at the wavelength λ_1 saturates SOA1 and induces a gain and refractive index change. A probe signal P_{cnv} at a new wavelength λ_2

---

[2] This factor is termed as a linewidth enhancement factor, because the spectral width Δf of a single-mode semiconductor laser is proportional to $(1+\alpha_H^2)$ [27].

[3] The word "interband" is used for the carrier dynamics between the conduction band and the valence band, while the word "intraband" is used for the carrier dynamics inside the conduction band or the valence band.

experiences a phase shift. If the phase shift equals π, the probe signal is redirected from the sum output port to the difference output port P_Δ.

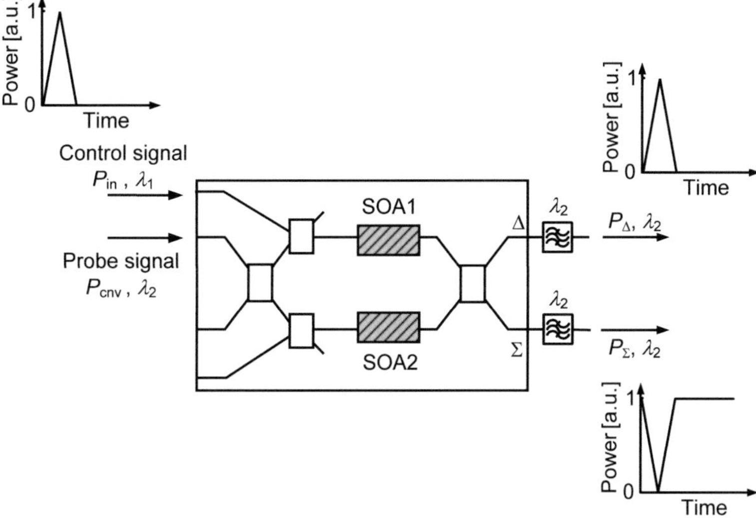

Fig. 1.3. All-optical wavelength converter based on a Mach-Zehnder interferometer configuration comprising two equally biased SOAs.

The MZI scheme can be operated in a differential mode. The differential operation scheme of a MZI allows for short switching windows, which are only determined by the fast refractive index change from carrier depletion and not affected by the slow carrier recovery time of the SOAs. The pattern effect related to the slow carrier recovery time is then mitigated. As shown in [69], the switching window with maximum transmission has been found to be limited to about 1.5 ps.

The SOAs on the MZI arms need to be optimized to obtain an optimum extinction ratio between different states with and without a control signal. Because the gain in SOA1 (see Fig. 1.3) changes while the gain of SOA2 (see Fig. 1.3) remains unchanged, the "off" state at the output port P_Δ is not optimal. Thus, to improve the extinction ratio, the current supplying SOA2 needs to be reduced. Further, an additional phase-shifting is required to compensate the undesired phase shifts that occur when differently biasing the SOAs. This method was presented in the reference [38].

Two SOAs are used in the MZI. This increases the expense. Next, schemes using a single SOA are introduced.

SOA in Sagnac Interferometer

Fig. 1.4 depicts the setup of a Sagnac loop with an asymmetrically placed SOA. It consists of a directional coupler C1 (mostly a 50:50 coupler[4]) with two branches connected to a loop. The loop contains a polarization control (PC2) and an SOA. A probe signal P_{cnv} at the wavelength λ_2 is split in the coupler into two parts. As shown in Fig. 1.4, the part traveling in clockwise direction is indexed with a superscript "c", P_{cnv}^c. Another part traveling in counterclockwise direction is indexed with a superscript "cc", P_{cnv}^{cc}. A control signal P_{in} at the wavelength λ_1 is launched into the loop by another coupler C2, see Fig. 1.4. The signal P_{in} modulates the gain and the refractive index of the SOA. Both signal, P_{cnv}^c and P_{cnv}^{cc}, experience the gain modulation and phase modulations. While the SOA output signals' shape is distorted due to the slow carrier recovery, the Sagnac interferometer is used to differentiate the SOA's output. The differentiation works as follows. As shown in Fig. 1.4, the SOA is asymmetrically placed in the loop. With respect to the center of the SOA, the modulated co-propagating signal P_{cnv}^c reaches the coupler C1 first and the modulated counter-propagating signal P_{cnv}^{cc} arrives later by a time ΔT. At the transmission port of the coupler C1, the phase difference between P_{cnv}^c and P_{cnv}^{cc} is π. This is because from the input to the remission the probe signal crosses the coupler C1 twice, while each cross coupling is accompanied with a $\pi/2$ phase shift. The phase difference in this time window ΔT then leads to the differentiation between two signals P_{cnv}^c and P_{cnv}^{cc}, and restores the shape of the incident signal.

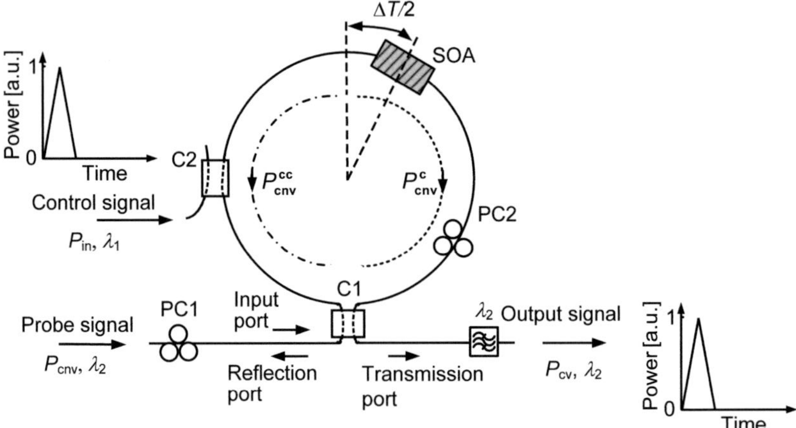

Fig. 1.4. All-optical wavelength converter based on a Sagnac loop with an asymmetrically placed SOA. PC1 and PC2 are polarization controllers before and in the Sagnac loop respectively.

---

[4] A 50:50 coupler has 50:50 power splitting ratios for the bar and cross transmissions. It is also called as a 3 dB coupler.

SOA followed by a delay interferometer

The wavelength conversion scheme comprising an SOA followed by a delay interferometer (DI) is shown in Fig. 1.5. The longer arm of the DI provides a time delay of ΔT and the shorter contains a phase shifter. The control signal P_{in} at the wavelength λ_1 modulates the gain of the SOA and thereby the phase of the probe signal P_{cnv} at wavelength λ_2. The DI is then used to transform the phase-modulation into amplitude modulation as follows. As the modulated signal P_{inv} is guided into the DI, it is split into two paths. The duplicate on the shorter DI arm (lower arm in Fig. 1.5) reaches the coupler first. If the induced phase in the P_{inv} signal is $\sim \pi$, the phase difference between the signals on the DI arms then opens the switching window for the output port P_{cv}. A time ΔT later, when the second duplicate on the longer (upper) DI arm reaches the coupler, the phase difference is reset and the switching window closes. Adjusting the phase shifter on the shorter DI arm can also control and optimize the output signal after the DI. Details of the operation of this device will be given later in Chapter 4.

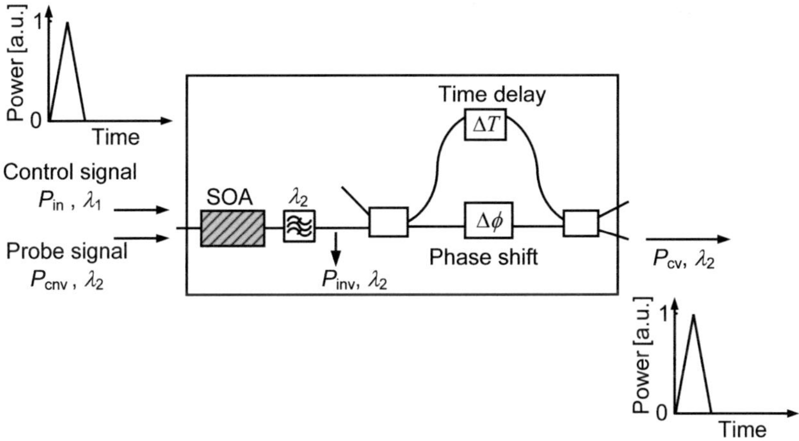

Fig. 1.5. All-optical wavelength converter based on an SOA followed by a delay interferometer.

SOA followed by an optical filter

Recently, a new approach of a wavelength converter with an SOA followed by an optical filter was introduced and discussed, Fig. 1.6. For all-optical wavelength conversion, the input data signal P_{in} and the probe (cw) light P_{cnv} are launched into the SOA. In the SOA, the input data signal encodes the signal information by means of cross-gain and cross-phase modulation onto the cw signal. As a result, we obtain after the SOA an inverted signal P_{inv} at the wavelength of the cw light. The purpose of the subsequent passive optical filter with an appropriate amplitude and phase response is to reformat the signal P_{inv} into a new output-signal P_{cv} with a predetermined shape.

Several filters have been proposed. These are the pulse-reformatting optical filter (PROF) [45] and [46], the red-shift optical filter (RSOF) [44], the blue-shift optical filter (BSOF) [44],[49], [50] and [62], and a delay interferometer (DI) filter scheme [41], [47].

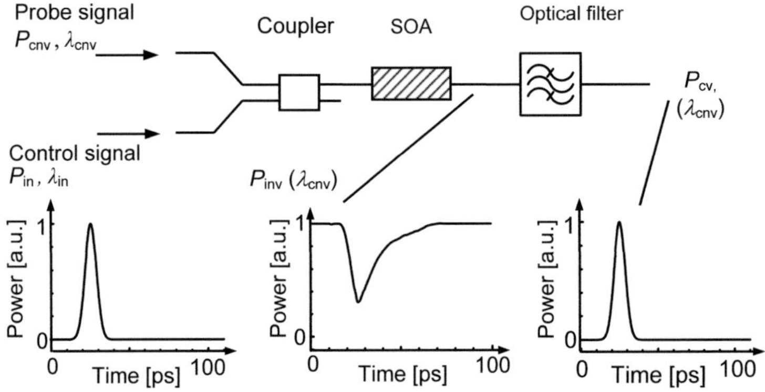

Fig. 1.6. All-optical wavelength converter based on an SOA followed by an optical filter.

The PROF scheme can be understood as follows. As seen Fig. 1.6 the probe signal P_{cnv} is cross-phase modulated by the control signal P_{in}, the P_{inv} signal after the SOA comprises a leading red-chirped (negative frequency variation with respect to time) component followed by a trailing blue-chirped (negative frequency variation with respect to time) component due to SOA carrier recovery. The P_{inv} signal also has a cw tone from the P_{cnv} signal. The PROF suppresses the cw tone and splits off the red and blue-chirped spectral components, which are subsequently recombined. This leads to a beating between the two signals that results in a strong and narrow converted pulse if the temporal delay between the two pulses is adapted correctly. A schematic of the filter is depicted in Fig. 1.7(a) and the passband of the filter around the cw frequency f_{cw} is shown in the right part of Fig. 1.7(a).

The RSOF and BSOF schemes are similar to the PROF scheme, as shown in Fig. 1.7(b) and (c). Yet, either the red-chirped or the blue-chirped spectral components from the converted signals are selected by the respective filter, while the blue-chirped or the red-chirped spectral components are blocked as indicated by a symbol "X" in Fig. 1.7(b) and (c). Indeed, the DI filter in Fig. 1.5 can also be used to perform regenerative wavelength conversion, either as a red-shift optical filter [45] or a blue-shift optical filter [49], [50] and [63]. With a proper choice of the BSOF's or RSOF's shape, the output signal P_{cv} shows a predetermined pulse shape. Details of the PROF, BSOF, and RSOF schemes are given in Chapter 4.

The PROF, RSOF, and BSOF schemes exploit the fast chirp effects in the converted signal after the SOA, and are successful in performing high-speed wavelength conversion [44], [45], [46], [49], [50] and [62]. However, so far it is not clear, if these schemes with optical filters can successfully overcome pattern effects at highest speed. In this work, we will investigate how to mitigate the pattern effect by using respective filters.

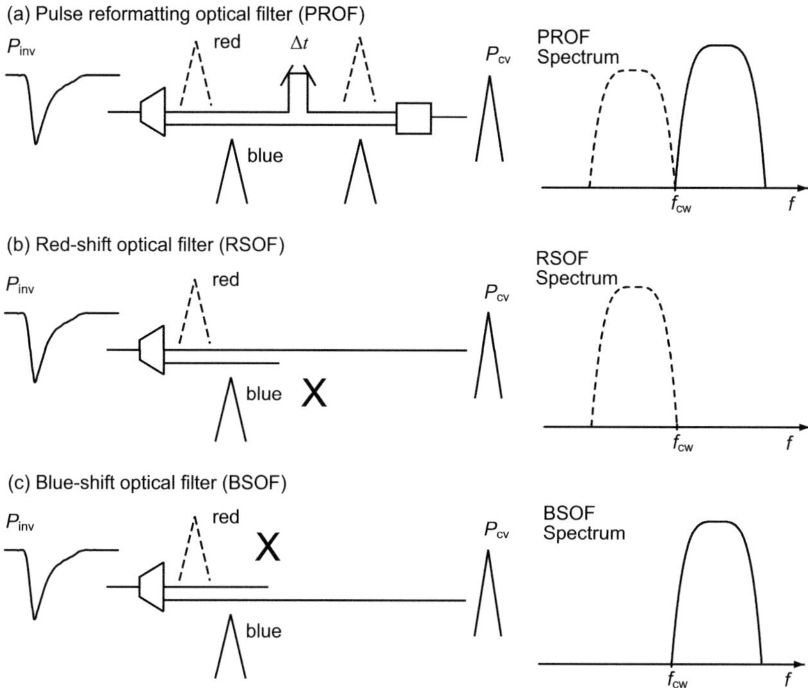

Fig. 1.7. Three different optical filter concepts for performing wavelength conversion. In the PROF scheme (a), the red- and blue-chirped spectral components are selected and combined with a proper temporal delay Δt between two paths. In the RSOF scheme (b), the red-chirped spectral component is selected and the blue-chirped spectral component is blocked (indicated with a symbol "X"). In the BSOF scheme (c), the blue-chirped spectral component is selected and the red-chirped spectral component is blocked.

1.2 Achievements in this Work

- **An SOA model has been established.** The SOA model is based on a rather simple set of rate equations for inter- and intraband carrier dynamics. The intraband carrier dynamics are CH and SHB. The rate equations are simplified by assuming charge neutrality and neglecting the longitudinal and transverse carrier diffusions. The traveling wave equations are used to describe the pulse propagation through the SOA.

  The SOA model has been validated by simulating the (power) gain and phase responses of an SOA in a wavelength conversion experiment. Proper phase response is modeled by using individual linewidth enhancement factors, also α-factors, for individual carrier dynamics. Especially, a novel parameterization for the α-factor corresponding interband carrier dynamics was introduced. Inclusion of this model leads to excellent simulation results in agreement with 160 Gbit/s experiment. This experimentally validated model is then implemented in an optical communication system simulator, namely OptSim from RSoft [94].

- **An effective α-factor was introduced and its dynamics in a cross-gain modulation experiment was investigated.** The effective α-factor relates the total phase change with the measured gain change of the output signal out of an SOA. In the experiment and simulation, this effective α-factor varies strongly with time, different to a constant α-factor assumed in [27]. It even takes on negative values for short periods of time. As a consequence, XPM effects usually lag behind XGM effects by several picoseconds. This delay has important consequences for operation of all-optical devices based on SOAs.

- **Return-to-zero (RZ) to non-return-to-zero (NRZ) wavelength conversion at 160 Gbit/s was achieved by using an SOA followed by a delay interferometer filter.** Experimental results from this scheme, as shown in Fig. 1.5, fit to the simulation results very well and to a degree not possible before.

- **Wavelength conversion with an SOA followed by a pulse reformatting optical filter was investigated.** In this work, we will show that the PROF scheme indeed and effectively mitigates SOA patterning. An experimental implementation at 40 Gbit/s shows a signal quality factor improvement of 7.9 dB and 4.8 dB if compared to a blue- or red-shifted optical filter assisted wavelength converter scheme, respectively. This technique is also supported by simulation.

- **Wavelength conversion in an SOA-based Sagnac loop was investigated, especially, for non-return-to-zero (NRZ) modulation.** A few NRZ wavelength conversions with a SOA-based Sagnac loop have been successfully demonstrated at 10 Gbit/s [87] and at 42.7 Gbit/s [30]. However, the influence of the SOA properties, especially the SOA carrier recovery time, on the performance of the wavelength conversion has been neither discussed nor been well understood. In this work, we discuss how operation speed and SOA dynamics are related. We found that SOA recovery times can be neither too fast nor too slow for a particular operation speed. An optimal carrier recovery time is between two and three times of one bit slot duration. We also found that shorter SOAs are preferred in this wavelength conversion scheme.

- **Wavelength conversion for differential phase-shift keying (DPSK) modulation was demonstrated in an SOA based MZI scheme.** Its regenerative properties as well as the cascadability has been discussed, where the cascaded configuration is built up with a number of identical DPSK wavelength converters.

2 Semiconductor Optical Amplifier: Theory and Modeling

This Chapter describes the light propagation through the semiconductor optical amplifier (SOA) and the interaction between the light and the gain medium. While the light propagation can be modeled through the wave equations, the interaction between the gain medium and the input light is modeled by rate equations accounting for carrier dynamics in the gain medium. The carrier dynamics include either interband or intraband processes in the gain medium. Interband processes comprise spontaneous recombination, stimulated radiative recombination and carrier transport, whereas intraband processes include carrier heating (CH) and spectral-hole burning (SHB). The carrier dynamics includes also two-photon absorption (TPA) process. These dynamic processes affect both gain and refractive index of an SOA. In addition, to describe the phase dynamics of the optical signal, a novel parameterization of the linewidth enhancement factor, i.e. α-factor, is introduced. To solve the equations, a multi-section numerical algorithm is implemented. The basic characteristics of the SOA, e.g. gain characteristics and spatial carrier distribution inside the SOA, are simulated and shown in this Chapter.

2.1 Pulse Propagation and Gain Saturation

2.1.1 Pulse Propagation and Amplification

Since the optical pulse also experiences power amplification along the SOA, the material property of the SOA needs to be discussed first. In this work, the material in the SOA is assumed to be isotropic (this is a valid assumption also in most practical cases). For such a material, as discussed in [2], [22], the medium response to the electric field can be described by a medium susceptibility χ, since the higher order susceptibilities are usually weak in the semiconductor material. The complex relative dielectric constant $\bar{\varepsilon}$ and the complex refractive index $\bar{n}$ are given by:

$$\bar{\varepsilon}(\omega) = \varepsilon_r(\omega) - j\varepsilon_{ri}(\omega) = 1 + \chi(\omega) = 1 + \chi_r(\omega) + j\chi_i(\omega),\qquad(2.1)$$

$$\bar{n}(\omega) = n_r(\omega) - jn_i(\omega),\qquad(2.2)$$

$$\bar{\varepsilon}(\omega) = \bar{n}^2(\omega),\qquad(2.3)$$

where ε_r, χ_r, n_r and ε_{ri}, χ_i, n_i are the real and imaginary parts of the dielectric constant, the medium susceptibility and the refractive index, respectively. $k_0 = 2\pi f/c_0$ is the wavenumber in vacuum.

In an isotropic SOA, we can use a scalar wave equation to describe the pulse propagation through an SOA. In the Cartesian coordinate system, we consider a linearly polarized plane wave propagating along the z-axis, shown in Fig. 1.1, while the x-y plane is perpendicular to the propagation direction. So by choosing the transverse-electric (TE) orientation of polarization parallel to the x-axis, we only consider the E_x component of the electric field vector $\vec{E}$.

We also assume that the spectral range of interest is sufficiently narrow around the optical carrier (angular) frequency $\omega_0 = 2\pi f_0$. The dielectric constant $\overline{\varepsilon}$ and also the complex refractive index $\overline{n}$ in this range are constant and can be represented by their values at frequency ω_0. Note that, by taking account of the waveguide structure and the material property of the SOA, the dielectric constant $\overline{\varepsilon}(\omega_0)$ and the refractive index $n_r(\omega_0)$ are effective values. Thus, the scalar wave equation is

$$\nabla^2 E(t) - \frac{\overline{\varepsilon}(\omega_0)}{c_0^2} \frac{\partial^2 E}{\partial t^2} = 0 , \tag{2.4}$$

where the x-subscript for the electric field is dropped.

For fields passing through an SOA, it is convenient as in [1] and [2] to write the dielectric function $\overline{\varepsilon}$ in terms of an effective refractive index n_r and an effective susceptibility χ_{eff}, which represents the contribution from the charge carriers inside the active region of the amplifier. Naturally, the susceptibility χ_{eff} is a function of the carrier density N. Therefore, the susceptibility χ_{eff} is further expressed by the power gain g and the change of the refractive index Δn_r, which are both functions of the carrier density N. In total, we have

$$\overline{\varepsilon} = n_r^2 + \chi_{eff}(N) \approx n_r^2 + \frac{n_r}{k_0}[2k_0 \Delta n_r(N) + jg(N)] , \tag{2.5}$$

$$\overline{n} = n_r + \Delta n_r(N) + j\frac{g(N)}{2k_0} . \tag{2.6}$$

Note that the high order terms, namely $(\Delta n_r(N))^2$, $\left(\frac{g(N)}{2k_0}\right)^2$ and $\Delta n_r(N)\frac{g(N)}{2k_0}$, are neglected in Eq. (2.5), since they are practically small. From the power gain $g(N)$ determined by the present carrier density N, a power gain change Δg can be calculated with respect to a reference value g_{ref} at a stationary state

$$\Delta g(N) = g(N) - g_{ref} . \tag{2.7}$$

If only one optical signal propagates through the SOA, as discussed in [1], the value at the stationary state g_{ref} is then 0 and $\Delta g(N) = g(N)$. In the case of an XPM, another control signal is introduced into the SOA and induces the gain and phase changes on the probe signal. The reference value g_{ref} in Eq. (2.7) is then the stationary state value before the incidence of the control signal.

In reality, the refractive index and the material gain or absorption are related by the Kramers-Kronig (KK) integral expressions [36] and [37]. Phenomenologically, a linewidth enhancement factor α_H, also called Henry factor [27] is used to relate the changes of the refractive index Δn_r and the power gain Δg [1] and [27]. The linewidth enhancement factor α_H is defined as the ratio between the real and the imaginary part of $\overline{n}$:

$$\alpha_H = \frac{\partial n_r / \partial N}{\partial n_i / \partial N} = -2k_0 \frac{\partial n_r / \partial N}{\partial g / \partial N}, \text{ and } \Delta n_r(N) \approx -\frac{\alpha_H}{2k_0} \Delta g(N). \tag{2.8}$$

Note that the second equation in (2.8) is an approximation by replacing the derivatives with the respective differences. The alpha-factor α_H is then dependent on the wavelength, the current density, and the material used for the SOA's. In most publications [1], [27] and [57], the α_H-factor of an SOA is assumed to be constant. This simplification indeed is valid only over a limited spectral range and for sufficiently small carrier density modulations in the SOA. Still, this concept of using the α_H-factor rather than the KK relations has shown to be useful for performing fast and efficient device simulations [2].

Eq. (2.4) and (2.5) can provide the description of the propagation of optical pulses in semiconductor optical amplifier. To simplify the analysis of the light propagation in the SOA, a few practical assumptions can be made:

- Only a single waveguide mode exists in the active region of the traveling-wave amplifier. The linearly polarized electrical field remains linearly polarized during propagation.

- The bias current is assumed to pass through the active region and not through the surrounding region. Furthermore a uniform distribution of bias current across the active region width is assumed.

- For the moment, the carrier distribution inside the SOA is assumed in quasi-equilibrium. The quasi-equilibrium means the equilibration of the carriers within their bands, namely CB and VB, but not among bands. The quasi-equilibrium is characterized by the quasi-Fermi levels W_{fc} and W_{fv} in the CB and the VB, respectively. The quasi-equilibrium distribution occurs when an external voltage is applied to the device of interest, or by applying a sufficient strong optical field.

- From the above assumption, the intraband relaxation effects, which would modify the value of the α_H-factor, are not taken into account. Later in this work, the variation of the α-factor will be discussed in details.

In Cartesian coordinates, we express the traveling electric field inside the amplifier as:

$$E(x,y,z,t) = \frac{1}{2}\vec{e}_x \left[F(x,y)\exp(j\omega_0 t)A(z,t)\exp(-jk_0 n_r z) + c.c. \right]. \tag{2.9}$$

The term $\vec{e}_x$ is the polarization unit vector of the transverse-electric (TE) field. The term $F(x,y)$ is the mode distribution. With an assumption of a plane wave, $F(z,t)$ is a real value and normalized as $\int_{-\infty}^{+\infty} |F(x,y)|^2 \, \mathrm{d}x\mathrm{d}y = 1$. The term $A(z,t)$ is the slowly-varying envelope of the traveling wave[5]. The variable z, also the coordinate, lies along the propagation direction through the SOA with its origin at the input facet. The term $c.c.$ indicates the complex conju-

---

[5] The slowly-varying envelope approximation is valid with respect to the time scale of the optical carrier f_0, so that $\left| \partial^2 [A(z,t)]/\partial z^2 \right| << \left| k_0 \partial [A(z,t)]/\partial z \right|$.

gate. For a causal system, the part $F(x,y)\exp(j\omega_0 t)A(z,t)\exp(-jk_0 n_r z)$ is an analytic signal. This analytic signal part alone is sufficient to describe the optical field. Thus, the conjugate complex part $c.c.$ is dropped away in the following.

The envelope function $A(z,t)$ is related to the signal power $P(z,t)$ and the phase shift $\Delta\varphi(z,t)$ as

$$A(z,t) = \sqrt{P(z,t)}\, e^{j\Delta\varphi(z,t)}. \tag{2.10}$$

In MKS system of units, the signal power P is further related to the photon density S as

$$P = h f v_g A_e S. \tag{2.11}$$

h is the Planck constant. v_g is the group velocity. A_e is the effective area calculated from the cross section area A of the active region by $A_e = A / \Gamma$, where Γ is the field confinement factor. This field confinement factor Γ describes the fractional contribution of the mode in the active region. For an active region with a width w and a height d (The cross section of the active region has then an area $A = wd$),

$$\Gamma = \frac{\int_0^w \int_0^d |F(x,y)|^2\, dxdy}{\int\int_{-\infty}^{+\infty} |F(x,y)|^2\, dxdy}. \tag{2.12}$$

The phase shift $\Delta\varphi$ is the difference between the instantaneous phase φ with respect to a reference value φ_{ref} at stationary state. From Eq. (2.6) and (2.9), the phase shift $\Delta\varphi$ at a position z can be calculated from the change of the refractive index Δn_r.

$$\Delta\varphi(z,t) = \varphi(z,t) - \varphi_{\text{ref}} = -k_0 \int_0^z \Delta n_r \left(N(z',t)\right) dz'. \tag{2.13}$$

The reference value φ_{ref} at stationary state can be assumed to be 0 for the sake of simplicity. As seen in Eq. (2.13), the change of the refractive index is dependent on the carrier concentration N and thus implicitly on the incident signal power P, since the incident power also modifies the carrier concentration in the active region.

Substituting the electrical field from Eq. (2.9) in the wave equation (2.4) and integrating over the transverse dimensions (x, y) will lead to one equation for the envelope function $A(z,t)$, [1] and [2],

$$\frac{\partial A(z,t)}{\partial z} + \frac{1}{v_g}\frac{\partial A(z,t)}{\partial t} = \frac{1}{2}[j\alpha_H \Delta g(z,t) + g(z,t)]A(z,t). \tag{2.14}$$

The terms g and Δg are also dependent on position and time, due to the interaction between the gain medium and the optical input signal. Note that to derive Eq. (2.14) the high-order derivatives of $A(z,t)$ over the position z and the time t are neglected[6]. Details of the derivation of Eq. (2.14) can be found in [2].

---

[6]Neglecting the high-order derivatives of $A(z,t)$ over the time t means that the group velocity dispersion is not taken into account. Including the croup velocity dispersion can be referred to in [28].

The active medium of an SOA provides the gain, which is determined by the carrier density inside the active region. This gain coefficient is thus given as a material gain g_m. However, only a fraction of the mode resides in the active region. Thus, taking into account the field confinement and the internal loss α_{int}, the power amplification for a mode is given by a modal gain g,

$$g(N(z,t)) = \Gamma g_m(N(z,t)) - \alpha_{int}(z,t) . \tag{2.15}$$

For the material gain and the material loss, in this work, there are a few assumptions:

- The SOA modal gain is polarization independent. This can be obtained by optimizing the structure of the active layer in the SOA [26].

- The transverse variations in the carrier density are negligible. This assumption is practically valid because of the narrow active region of SOA.

2.1.2 Gain Saturation

When an ultra-short but strong optical pulse propagates through the SOA, the intraband effects take place for a short period of time. The intraband effects considered in this work are spectral hole burning (SHB), carrier heating (CH), and carrier cooling. These intraband effects also modify the carrier distributions within the band, while the interband effect (e.g. stimulated emission) modifies the carrier distributions in both conduction and valance bands.

The temporal evolution of the carrier distribution, when an ultra-short optical pulse is injected in an SOA, is qualitatively illustrated in terms of the electron density versus the electron energy in the CB in Fig. 2.1.

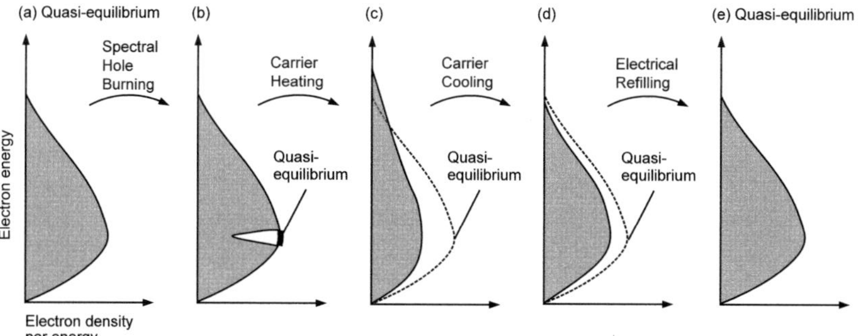

Fig. 2.1. Evolution of the spectral carrier density distribution versus energy in the conduction band, when an optical pulse is introduced in the SOA. The areas under the solid lines are the local carrier concentrations in the conduction band.

Fig. 2.1(a) shows the carrier distribution in the conduction band at quasi-equilibrium, which means the equilibration of the carriers within the conduction band (CB) but not among the CB and valance band. As an ultrashort optical pulse is injected in the SOA, the induced emission not only offers the power amplification for the input signal, but also causes a spectral hole in the carrier distribution, seen in Fig. 2.1(b). Obviously, the carrier distribution in Fig. 2.1(b) is not at quasi-equilibrium described by a Fermi distribution with the help of a quasi-Fermi level. The Fermi distribution at quasi-equilibrium is restored via intraband carrier scattering, as shown in Fig. 2.1 (c). The parameter to characterize the carrier scattering process is called SHB relaxation time constant. Now, the average carrier temperature is now higher than that of the lattice. In fact, the carrier depletion takes place mostly for the "cooler" carriers near the band edge, and therefore the average temperature of the carrier increases. In addition, free carrier absorption (FCA) effect also contributes to the increase of carrier temperature [57] and [86], where a free carrier absorbs a photon and jumps to a higher energy level. After the optical pulse leaves the SOA, the hot carriers release their excess energy via phonon emission. The time constant associated with this hot-carrier relaxation process is denoted as carrier-heating relaxation time constant or as temperature relaxation time constant. As the carriers cool down, the carrier distribution evolves to the Fermi distribution characterized by the lattice temperature, seen in Fig. 2.1(d). However, the total carrier number is less than that in the stationary state in Fig. 2.1(a). Then, the carriers are injected from external electrical circuit and refilled in the respective bands. Finally, the carrier distribution recovers to the quasi-equilibrium state, seen in Fig. 2.1(e).

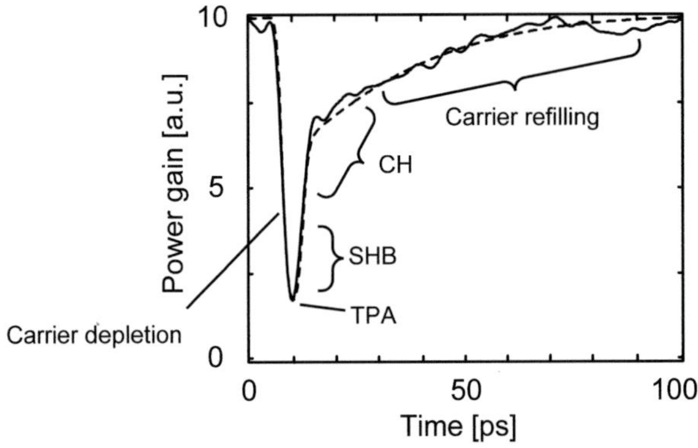

Fig. 2.2. Evolution of the power gain in a cross-gain modulation experiment. Solid and dashed lines are the measurement and simulation results, respectively. TPA: two-photon absorption; SHB: spectral hole burning; CH: carrier heating. From [82].

The temporal evolution of the carrier distribution also determines the temporal evolution of the material gain. Concretely, the carrier distribution in Fig. 2.1(a), (d) or (e) determines the material gain at quasi-equilibrium, while the non-equilibrium carrier distributions according to SHB and CH in Fig. 2.1 (b)-(c) lead to so called *nonlinear gain compression*. Practically, the temporal evolution of the power gain can be seen in a cross gain modulation (XGM) experiment, where the input pulse width is shorter or at least comparable to the intraband scattering times. Typical XGM responses from a measurement and a simulation [82] are shown as solid and dashed lines in Fig. 2.2. Details of this experiment will be given in Chapter 3.

The evolution of the gain recovery in Fig. 2.2 indicates the dominant nonlinear effects in the respective regimes. At first, the input pulse depletes the carriers. At the pulse peak power, the carrier depletion reaches its maximum and the power gain is the lowest. Afterwards, the carrier recovers. We now discuss the carrier recovery processes separately. First, if enough photons are present in the SOA, two-photon absorption (TPA) may take place, i.e. two photons are simultaneously absorbed and an electron is transferred from the VB to the CB. TPA is considered to be an instantaneous process as discussed in [26], [54], [58]. So, in Fig. 2.2, TPA can be seen at the lowest gain point. On the next hundred femtoseconds time scale after the lowest gain point, see Fig. 2.2, the dominant nonlinear effect is SHB. Afterwards, as discussed in [24] and [54], CH effect is dominant for a time scale of ~1 ps, see Fig. 2.2, to reestablish the Fermi carrier distribution in the CB and VB. The last time regime in Fig. 2.2 is the carrier refilling (recovery) process, which may take a time scale from some tens of picoseconds to a few nanoseconds, depending on the dimension of the active region, the material used in the active region and the SOA operation conditions.

If the input pulse width is much longer than the intraband scattering times, the CH and SHB effects can be observed as nonlinear gain compressions at stationary state (e.g. Fig. 2.1 (d)) with the instantaneous value of the field, [1], [57], [70], [86]. In this case, the nonlinear gain compressions are included phenomenologically in the material gain g_m for a stationary photon density S as[7]

$$g_m = \frac{\overline{g}_m}{1 + \varepsilon_{tot} S} \, . \tag{2.16}$$

The term $\overline{g}_m$ is the material gain without the gain compression, and ε_{tot} is the total nonlinear gain compression factor. It should be noticed that Eq. (2.16) may hide the fact that the nonlinear gain compression is determined by the carrier density not the photon density directly. As discussed in [70], the gain compression expression $\varepsilon_{tot} S$ in Eq. (2.16) is actually connected to a stationary carrier density N above the transparency carrier density N_0. The transparency carrier density N_0 is the carrier density level, where the material gain is 0 and the SOA is then a transparent device for the input signal.

The nonlinear gain compression factor ε_{tot} includes the contributions due to CH and SHB [57]

$$\varepsilon_{tot} = \sum_{\beta=c,v} \left(\varepsilon_{CH,\beta} + \varepsilon_{SHB,\beta} \right), \tag{2.17}$$

where the subscript β indicates the conduction band ($\beta = c$) or valence band ($\beta = v$). Details of the nonlinear gain compression factors are given in Appendix A.

In ultrafast SOA-based all-optical wavelength converters, the input pulse width becomes comparable to the intraband scattering times. In this case, the nonlinear gain compressions

---

[7]The gain compression in [86] is expressed as $g_m = \overline{g}_m (1 - \varepsilon_{tot} S)$. This expression is equivalent to Eq. (2.16) for $\varepsilon_{tot} S \ll 1$.

determined by using Eq. (2.16) are not proper any more. To model the material gain accurately, not only the carrier dynamics in the quasi-equilibrium has to be included, but also the dynamics of CH and SHB. The description will be given in the following.

2.2 SOA Modeling

In the section, we discuss the SOA modeling used in this work. In this model, we describe the carrier dynamics in the active region of the amplifier by using a rather simple set of rate equations for the carrier density and nonlinear gain compression terms related to CH and SHB, respectively. As given in [2], [4], [57], these rate equations have been derived from semi-classical density matrix equations by adiabatical elimination of the interband polarization (off-diagonal elements of the density matrix [8][8]) of the carrier. The derivation of the rate equations can be found in the literatures mentioned above. Thus, it is not repeated here.

The rate equation for the carrier density is important, since the carrier density inside the active region determines the material gain in quasi-equilibrium. To calculate the material gain coefficient, we implement a parameterization for SOAs with InGaAsP/InP material from [40], which models the material gain coefficient as a polynomial function of the total carrier density and the wavelength. This parameterization has been validated experimentally and can reduce the calculation time efficiently.

To take into account the bidirectional propagations, we extend the definition of the traveling electric field from Eq. (2.9) as

$$E(x,y,z,t) = \frac{1}{2}\vec{e}_x \{F(x,y)\exp(j\omega_0 t)$$
$$\times\left[A^+(z,t)\exp(-jk_0 n_r z) + A^-(z,t)\exp(jk_0 n_r z)\right]+c.c.\} ,$$

(2.18)

where $A^+(z,t)$ and $A^-(z,t)$ are envelope functions of the forward and backward traveling waves, distinguished by the superscripts "$+$" and "$-$". The forward-wave travels from the input facet at position $z=0$ to the output facet at position $z=L$ in Fig. 1.1. From Eq. (2.10) and (2.11), the envelopes $A^{+(-)}(z,t)$ are related to the photon densities $S^{+(-)}$ as:

$$A^+(z,t) \sim \sqrt{S^+(z,t)}\, e^{j\Delta\varphi^+(z,t)} , \quad A^-(z,t) \sim \sqrt{S^-(z,t)}\, e^{j\Delta\varphi^-(z,t)} .$$

(2.19)

The terms $\Delta\varphi^+(z,t)$ and $\Delta\varphi^-(z,t)$ are the nonlinear phase shifts as the waves propagate to the position z inside the SOA. They are the differences between the instantaneous phases and the phases at the stationary state, following the definition in Eq. (2.13). The respective equations for the photon density and the phase shift will be given in this section.

2.2.1 Material Gain and Nonlinear Gain Compression

Contributions to the material gain coefficient g_m at a wavelength λ come from the "linear" material gain term $\bar{g}_m$, the nonlinear SHB gain compression term g_β and the nonlinear CH gain compression term $g_{T,\beta}$

---

[8] The off-diagonal terms are sometimes called the "coherences", because they describe superposition of states.

$$g_{\mathrm{m}}(\lambda, N) = \overline{g}_{\mathrm{m}}(\lambda, N) + \sum_{\beta=c,v} \left[g_{T,\beta} + g_{\beta} \right]. \tag{2.20}$$

The subscript $\beta = c,v$ refer to the CB and VB respectively. In the positive gain range $\overline{g}_{\mathrm{m}}(\lambda, N) > 0$, the gain compression terms g_{β} and g_{β} take a negative value. The input probe and control signals experience an extra gain reduction due to TPA in the active region which adds another term. More precisely, the various contributions are:

1) *Linear gain term* $\overline{g}_{\mathrm{m}}$ depends on the carrier densities in the CB and VB at the quasi-equilibrium and is evaluated at the lattice temperature T_L for a given carrier density. For practical calculations, this term needs to be parameterized. In this work, a parameterization for SOAs with InGaAsP/InP material is used. It is based on a polynomial model composed of a quadratic and a cubic function which depends on the carrier density N and wavelength λ [40],

$$\overline{g}_{\mathrm{m}}(\lambda, N) = c_N [\lambda - \lambda_z(N)]^2 + d_N [\lambda - \lambda_z(N)]^3,$$
$$c_N = 3\frac{g_p(N)}{[\lambda_z(N) - \lambda_p(N)]^2}, \, d_N = 2\frac{g_p(N)}{[\lambda_z(N) - \lambda_p(N)]^3}, \tag{2.21}$$
$$g_p(N) = a_0(N - N_0) + a_0 a_1 N_0 \exp(-N/N_0),$$
$$\lambda_p(N) = \lambda_{p,0} - b_0(N - N_0) - b_1(N - N_0)^2, \, \lambda_z(N) = \lambda_{z,0} - z_0(N - N_0).$$

At the peak wavelength λ_p, the linear material gain $\overline{g}_{\mathrm{m}}$ has the maximum value g_p. The dependence of λ_p on the carrier density is expressed by b_0. g_p is also dependent on the carrier density N through the differential gain constant a_0 around the transparency carrier density N_0. Parameters a_1 and b_1 are used to fit the measurement in [40]. λ_z is the carrier-density dependent bandgap wavelength, where $\lambda_{z,0}$ is the value of λ_z at the transparency carrier density N_0, and the constant z_0 defines the dependency of λ_z on the carrier density N. The material gain $\overline{g}_{\mathrm{m}}$ modeled for various wavelengths and various carrier densities is shown in Fig. 2.3. Seen in Fig. 2.3, increase of the carrier density moves the gain peak to a lower wavelength (high frequency). This is because at a low carrier density the low energy levels are first filled and the high energy levels are filled at a high carrier density.

Fig. 2.3 also shows a gain coefficient $\overline{g}'_{\mathrm{m}}$ for stimulated transitions from the CB to the VB, obtained by subtracting the stimulated absorption from the linear material gain $\overline{g}_{\mathrm{m}}$ [9]. In fact, the spontaneous transitions from the CB to the VB have the same spectral shape as the gain coefficient $\overline{g}'_{\mathrm{m}}$ [9] and [89]. As discussed in [2], [9] and [20], the gain coefficients $\overline{g}_{\mathrm{m}}$ and $\overline{g}'_{\mathrm{m}}$ are related by the inversion factor n_{sp},

$$n_{\mathrm{sp}} = \frac{1}{1 - \exp\left(\frac{hf - (W_{f_c} - W_{f_v})}{k_B T}\right)} = \frac{\overline{g}'_{\mathrm{m}}}{\overline{g}_{\mathrm{m}}}, \tag{2.22}$$

where k_B is the Boltzmann constant and T is the absolute temperature. In Fig. 2.3, the quasi-Fermi level separation $(W_{f_c} - W_{f_v})$ corresponds to the transparency wavelength point λ_0, where $\overline{g}_{\mathrm{m}} = 0$.

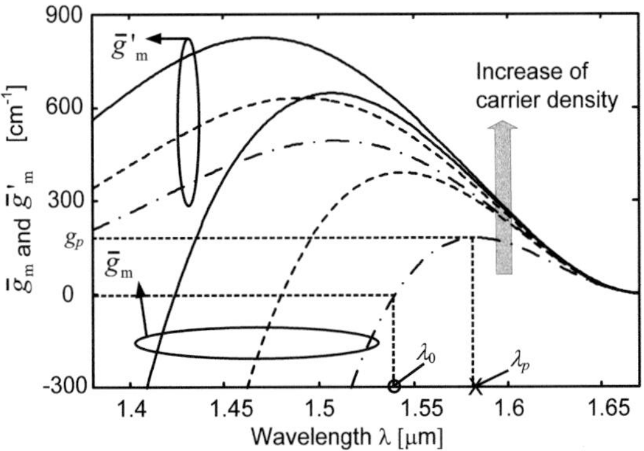

Fig. 2.3. Spectra of net material gain coefficient $\bar{g}_m$ and the gain coefficient $\bar{g}'_m$. They are connected via Eq. (2.22). At a given carrier density, the symbol "X" indicates the peak wavelength λ_p and the circle indicates the transparency wavelength λ_0, where $\bar{g}_m = 0$. Here the gain compression is not included and the value of all the parameters can be found in Appendix B.

2) *Gain compression terms $g_{T,\beta}$ and g_β* have a spectral dependency. We calculate its values at the peak gain wavelength λ_p and then bring in the spectral dependence through a weighting factor. For a pulse width comparable to the temperature relaxation time constants $\tau_{T,\beta}$ in the CB and VB (~1 ps) but much larger than the SHB relaxation time (~100 fs), the gain compression terms are calculated [57] with the respective gain compression factors $\varepsilon_{CH,\beta}$ and $\varepsilon_{SHB,\beta}$ for CH and SHB,

$$\frac{\partial g_{T,\beta}(\lambda_p)}{\partial t} = -\frac{g_{T,\beta}(\lambda_p)}{\tau_{T,\beta}}$$
$$-\frac{\varepsilon_{CH,\beta}}{\tau_{T,\beta}}\left(\sum_k g_{m,k}(s_k^{ASE,+} + s_k^{ASE,-}) + \sum_i g_{m,i}(S_i^+ + S_i^-)\right), \tag{2.23}$$

$$g_\beta(\lambda_p) = -\varepsilon_{SHB,\beta}\left(\sum_k g_{m,k}(s_k^{ASE,+} + s_k^{ASE,-}) + \sum_i g_{m,i}(S_i^+ + S_i^-)\right), \tag{2.24}$$

where $g_{m,i} \equiv g_m(\lambda_i)$ and $g_{m,k} \equiv g_m(\lambda_k)$. The index k extends over all the longitudinal modes due to amplified spontaneous emission (ASE) and s_k^{ASE} stands for the corresponding photon density. The index i extends over all the input signals and S_i stands for the photon density of the input signal. For instance, in the case of wavelength conversion shown in Fig. 1.2, the symbol i with "cnv" and "in" represents the probe signal and control signal. S_i is related to the signal power P_i of the input signal as in Eq. (2.19).

When reflecting facets are present, the spontaneously emitted noise will show the presence of longitudinal cavity modes[9] in the spectral range of interest, i.e. the optical bandwidth of the SOA. The longitudinal mode spacing of a cavity with a length L is given as

$$\Delta f = \frac{v_g}{2L}.$$
(2.25)

The mode frequency f_k^{ASE} of the k-th longitudinal mode due to ASE is given as in [9]

$$f_k^{\mathrm{ASE}} = f_c + \Delta f_c + k\Delta f, \quad \text{for} \quad k = 0,..., N^{\mathrm{ASE}} - 1,$$
(2.26)

where the cut-off frequency f_c is the frequency point, at which the modal gain falls to zero for increasing λ. Δf_c is a frequency offset used to match f_0^{ASE} to a longitudinal mode N^{ASE} is the number of the longitudinal modes within the linewidth of the SOA, i.e. the spontaneous emission spectrum as same as the curve of $\overline{g}'_\mathrm{m}$ in Fig. 2.3.

Eq. (2.24) is an expression derived from the corresponding rate equation in [57] (also given in Eq. (A.13) in Appendix A). This is done by assuming that the pulse width is much larger than the SHB relaxation time (~ 100 fs) and the SHB gain compression is at quasi-equilibrium. Under such an assumption, using Eq. (2.24) is sufficient to calculate the SHB gain compression.

The spectral dependence of $g_{T,\beta}$ and g_β is due to the spectral dependence of the nonlinear gain compression factors $\varepsilon_{\mathrm{CH},\beta}$ and $\varepsilon_{\mathrm{SHB},\beta}$ [57]. To calculate the gain compression $g_{T,\beta}$ and g_β for other wavelengths, rather than bringing in this dependence through various gain compression factors, we multiply the results from Eq. (2.23) and (2.24) with the weighting factors. For $g_\beta(\lambda)$, the weight factor is the ratio of $\overline{g}_\mathrm{m}(\lambda)$ to $\overline{g}_\mathrm{m}(\lambda_p)$. So $g_\beta(\lambda)$ is approximated by

$$g_\beta(\lambda) = g_\beta(\lambda_p)\frac{\overline{\overline{g}}_\mathrm{m}(\lambda)}{\overline{g}_\mathrm{m}(\lambda_p)}.$$
(2.27)

The spectral dependence of $g_{T,\beta}$ is similar to that of the gain coefficient $\overline{g}'_\mathrm{m}$, shown in Fig. 2.3. It can be obtained by multiplying $g_{T,\beta}(\lambda_p)$ from (2.23) with a normalization factor based on the inversion factor n_{sp} in Eq. (2.22)

$$g_{T,\beta}(\lambda) = g_{T,\beta}(\lambda_p)\frac{n_\mathrm{sp}(\lambda)}{n_\mathrm{sp}(\lambda_p)}.$$
(2.28)

The choice of the weighting factors for the respective compression terms ensures that $g_\beta(\lambda)$ for SHB has the inverse sign as the linear gain term $\overline{g}_\mathrm{m}(\lambda)$, while the gain compression due to CH does not [86].

---

[9]A longitudinal mode of a resonant cavity is a particular standing wave pattern formed by waves confined in the cavity. The longitudinal modes correspond to the wavelengths of the wave which are reinforced by constructive interference after many reflections from the cavity's reflecting surfaces. All other wavelengths undergo destructive interference and are suppressed.

Using Eq. (2.23), (2.24), (2.27) and (2.28), we can calculate the nonlinear gain compression of each mode of interest, e.g. the input signals and the ASE noise. The modeling results are illustrated in Fig. 2.4.

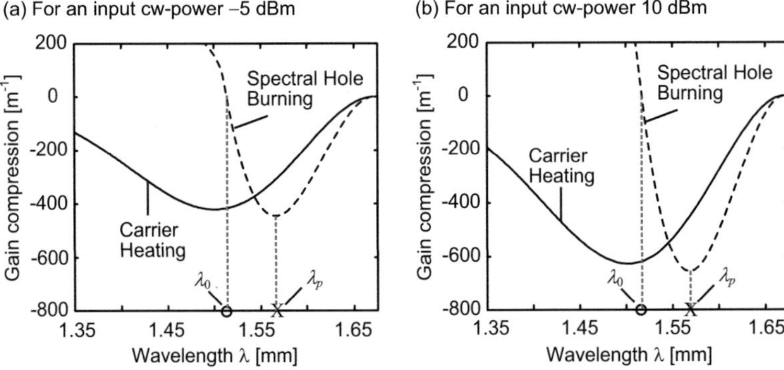

Fig. 2.4. Spectral dependence of gain compression $\sum_{\beta=c,v} g_\beta$ **(spectral hole burning) and** $\sum_{\beta=c,v} g_{T,\beta}$ **(carrier heating) at stationary state, where the subscript** $\beta = c, v$ **refer to the CB and VB respectively. The symbols "X" on the x-axis indicates the peak point** λ_p **of the material gain. The circles indicate the transparency wavelength** λ_0**, where** $\overline{g}_m = 0$**. Modeling is made based on [86]. Values of all the parameters can be found in Appendix B.**

3) Two Photon Absorption terms contributing to the gain compression are included in the gain terms of the input signals only - not in terms with ASE. In case of wavelength conversion, where there are one probe signal and one control signal, we only consider two types of two-photon absorption. The first case is the absorption of one photon from the probe signal and one photon from the control signal. The second case is the absorption of two photons from the control signal. As described in [29], [58], and [59], the TPA gain compression can be calculated empirically by using a TPA coefficient β_2 (in units of m^2):

$$g_{\text{TPA,cnv}} = -2\beta_2 S_{\text{in}}, \quad g_{\text{TPA,in}} = -\beta_2 S_{\text{in}}. \tag{2.29}$$

The factor 2 in the TPA gain compression for the probe signal represents the cross coupling between the probe signal and the control signal [58]. The slight dependence of TPA coefficient β_2 on material and wavelength [80] is neglected here, i.e., β_2 is assumed to be constant in the wavelength range of interest. Note that the TPA terms will be added phenomenologically to the gain terms of the input signals - but not to terms with ASE. The reason is that the ASE photon densities in the longitudinal modes are still weak, while TPA becomes only important at relative high photon densities [59].

4) Single Pass Gain and Other Parameters

From Eq. (2.20), (2.21),,(2.23) and (2.24), the material gain is implicitly dependent on the input power, since the input power changes the carrier density N in the active region. Practically, a single pass gain is defined as the power ratio between the SOA input and output signal. It is also implicitly dependent on the input power and can be calculated as

$$G_s(\lambda_i, N) = \exp\left[\int_0^L g_i(N)dz\right] = \exp\left[\int_0^L (\Gamma_i g_{m,i}(N) - \alpha_{int,i} + g_{TPA,i})dz\right], \quad (2.30)$$

where $g_i \equiv g(\lambda_i)$ is the net modal gain of the input signal indexed with a subscript i and $g_{m,i} \equiv g_m(\lambda_i)$, $\Gamma_i \equiv \Gamma(\lambda_i)$, $\alpha_{int,i} \equiv \alpha_{int}(\lambda_i)$ and $g_{TPA,i} \equiv g_{TPA}(\lambda_i)$. Indeed, the confinement factor Γ and the internal loss α_{int} are also modeled as functions of the carrier density N and wavelength λ.

A) *Confinement factor* Γ defined in Eq. (2.12) depends on the transversal mode profile. However, the exact calculation of the transversal mode profile consumes extraordinary computer time. Instead, a simple approximation given in [65] is used in this work. Also, the confinement factor Γ is modeled to be linearly dependent on the carrier density N and wavelength λ. The linear dependencies are expressed through the parameters $\left.\dfrac{\partial \Gamma}{\partial N}\right|_{N_{\Gamma_0},\lambda_{\Gamma_0}}$ and $\left.\dfrac{\partial \Gamma}{\partial \lambda}\right|_{N_{\Gamma_0},\lambda_{\Gamma_0}}$ at the fitting point with a wavelength λ_{Γ_0} and a carrier density N_{Γ_0},

$$\Gamma(N,\lambda) = \Gamma_0 + \left.\frac{\partial \Gamma}{\partial N}\right|_{N_{\Gamma_0},\lambda_{\Gamma_0}} (N - N_{\Gamma_0}) + \left.\frac{\partial \Gamma}{\partial \lambda}\right|_{N_{\Gamma_0},\lambda_{\Gamma_0}} (\lambda - \lambda_{\Gamma_0}). \quad (2.31)$$

Γ_0 is the value of Γ at the fitting point. The parameterization is shown in Fig. 2.5(a), and the fitted parameters are given in Appendix B.

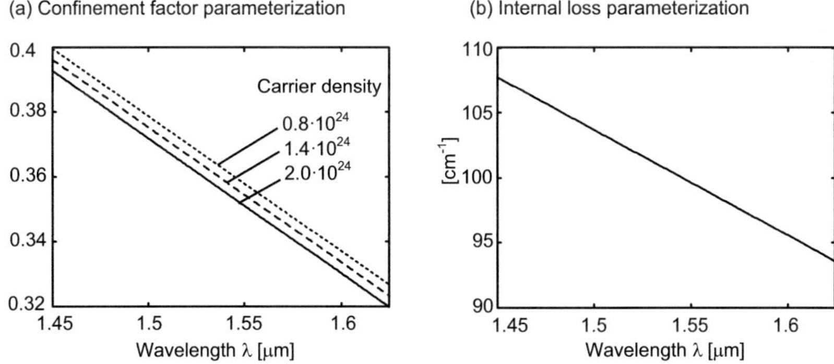

Fig. 2.5. Parameterizations of (a) the confinement factor Γ and (b) the internal loss coefficient α_{int}. Parameters are given in Appendix B.

B) *Internal loss* α_{int} is modeled to be linearly dependent on the wavelength λ, based on the measurement shown in [39],

$$\alpha_{int}(\lambda) = K_1(\lambda - \lambda_{p,0}) + \alpha_{int,0}, \quad (2.32)$$

where $\alpha_{int,0}$ is the value of α_{int} at peak wavelength $\lambda_{p,0}$ at transparency. This wavelength independent internal loss is due to scattering loss at the hetero-interface in the waveguide [72]. K_1 gives the dependence of α_{int} on the wavelength. The parameterization is shown in Fig. 2.5(b), and the fitted parameters are given in Appendix B.

2.2.2 Rate Equation for Carrier Density

The dynamics of the carrier distribution is obtained by solving a simple rate equation of the total carrier density N. As we assume charge neutrality, N accounts for both electrons and hole. In the following, we also neglect the longitudinal and transverse carrier diffusion and assume a uniform distribution of the bias current I_{bias} across the active region width. The volume of the active region is $V_{act} = AL$, where A and L are the cross section area and the length of the active region.

The carrier density at a position z inside of the SOA obeys the rate equation

$$\frac{\partial N(z,t)}{\partial t} = \frac{I_{bias}}{qV_{act}} - R_{tot} - 2v_g \sum_k g_{m,k}(s_k^{ASE,+} + s_k^{ASE,-}) - v_g \sum_i [(g_{m,i} + \frac{g_{TPA,i}}{\Gamma_i})(S_i^+ + S_i^-)], \quad (2.33)$$

where q is the elementary charge.

The first term on the right hand side of Eq. (2.33) describes the carrier injection from the applied current bias, the second term R_{tot} is the total spontaneous recombination rate, the third term is the stimulated emission caused by amplified spontaneous emission (ASE), and the fourth term is the stimulated emission caused by the input signal. The factor 2 before the third term takes into account the two mutually orthogonal polarizations (TE or TM). This factor does not appear before the fourth term, because we have assumed that the polarized field remains linearly polarized during propagation. The free-carrier generation due to TPA in the active region [58] is expressed via a gain compression term g_{TPA}. The product $\frac{g_{TPA}}{\Gamma}S_i$ in Eq. (2.33) is equivalent to a product $g_{TPA}P_i$, see Eq. (2.11), which tells that TPA takes place not only in the active region.

The spontaneous recombination rate R_{tot} in Eq. (2.33) is

$$R_{tot} = N/\tau_N = A_{SRH}N + B_{BB}N^2 + C_{Aug}N^3 + D_{leak}N^{5.5}, \quad (2.34)$$

where τ_N is the carrier lifetime. A_{SRH} is the Schockley-Read-Hall recombination coefficient due to traps in the semiconductor material [20], B_{BB} is the radiative band-to-band recombination coefficient, C_{Aug} is the Auger recombination coefficient [20], and D_{leak} takes into account the carrier leakage through the buried active region [35], while the exponent "5.5" was fit from the experiment.

2.2.3 Equations for Photon Density and Phase Shift

Equations for photon density of input signals

The propagation of the forward and backward traveling waves in Eq. (2.18) is now described by equations for the photon densities (i.e. signal power) and equations for the nonlinear phase changes. The reason to do so is that the nonlinear effects, e.g. band-filling, SHB and CH, influence the power amplification and nonlinear phase change differently. This difference will be discussed later in Chapter 3.

Following Eq. (2.14), one gets equations for the photon densities of the forward and backward traveling waves, $S^+(z,t)$ and $S^-(z,t)$,

$$\frac{\partial S_i^+(z,t)}{\partial z} + \frac{1}{v_g}\frac{\partial S_i^+(z,t)}{\partial t} = \left[(\Gamma_i g_{\mathrm{m},i} - \alpha_{\mathrm{int},i}) + g_{\mathrm{TPA},i}\right] S_i^+(z,t), \tag{2.35}$$

$$\frac{\partial S_i^-(z,t)}{\partial z} - \frac{1}{v_g}\frac{\partial S_i^-(z,t)}{\partial t} = -\left[(\Gamma_i g_{\mathrm{m},i} - \alpha_{\mathrm{int},i}) + g_{\mathrm{TPA},i}\right] S_i^-(z,t), \tag{2.36}$$

where the index i extends over all the input signals. Note that the material gain g_m from Eq. (2.20) is used, which includes the "linear" material gain term $\overline{g}_\mathrm{m}$, the nonlinear SHB gain compression term g_β and the nonlinear CH gain compression term $g_{T,\beta}$. In addition, the gain compression term due to TPA g_{TPA} is also added.

Equations for nonlinear phase shift of input signal

To properly model the evolution of phase shifts $\Delta\varphi^+(z,t)$ and $\Delta\varphi^-(z,t)$ defined in Eq. (2.19), the α-factor arising from the various effects must be included [57] and [65], incorporating the appropriate time constants. We first define individual α-factors associated with band-filling, CH, SHB and TPA. They are labeled with the subscripts N, T, SHB and TPA, respectively,

$$\alpha_N = -2k_0 \frac{\Delta n_N}{\Delta g_N}, \tag{2.37}$$

$$\alpha_{T,\beta} = -2k_0 \frac{\Delta n_{T,\beta}}{\Delta g_{T,\beta}}, \tag{2.38}$$

$$\alpha_{\mathrm{SHB},\beta} = -2k_0 \frac{\Delta n_\beta}{\Delta g_\beta}, \tag{2.39}$$

$$\alpha_{\mathrm{TPA}} = -2k_0 \frac{\Delta n_{\mathrm{TPA}}}{\Delta g_{\mathrm{TPA}}}. \tag{2.40}$$

The quantities $\Delta n_{N,(T,\beta),(\beta),(\mathrm{TPA})}$ and $\Delta g_{N,(T,\beta),(\beta),(\mathrm{TPA})}$ are the respective changes of the refractive index and of the gain due to band-filling, CH, SHB and TPA. Note that, for the sake of simplicity, we assume that α-factors associated with CH and SHB are same for the CB and the VB. So the subscripts "β" of $\alpha_{T,\beta}$ and $\alpha_{\mathrm{SHB},\beta}$ are dropped in the following.

The factors α_T, α_{SHB} and α_{TPA} are chosen empirically from our experiments, which will be shown in Chapter 3. The factor α_T, depends on the carrier energy inside the band [14] and varies especially near the band edge, where the depletion of "cooler" carriers takes place.

However, usually in the experiment, the signal photon energies influence carriers well inside the band. So, α_T, may be approximated by a constant. The spectral-hole burning (SHB) effect hardly contributes to the refractive index near the peak wavelength. This is because SHB happens symmetrically around the central wavelength, and the corresponding Kramers-Kronig integral will be essentially nearly anti-symmetric [57] and [60]. As a result, the integration over the whole spectrum is nearly zero and the change of the refractive index is then very small. This leads to $\alpha_{\mathrm{SHB}} \approx 0$. The TPA-induced factor α_{TPA} can be related to the conventional Kerr coefficient n_2 [60] and [80], where n_2 is assumed to be constant over the spectral range of interest, i.e. the linewidth of SOA.

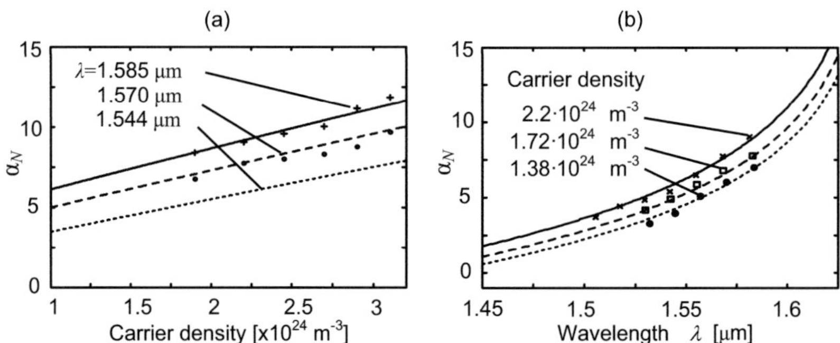

Fig. 2.6. Measured and parameterized values of the linewidth enhancement factor α_N as a function of (a) carrier density and (b) wavelength. The symbols represent the measured results [39]. The parameterized curves are plotted from Eq. (2.41) with the parameters given in Appendix B.

The factor α_N related to the band-filling needs more discussion. As there is no model that would cover both the spectral range of interest and the strong carrier density variation, we introduce a new parameterization for α_N. This parameterization is in agreement both with theory [66], [79], [85] and with the measurements [39]. The α_N introduced here depends both on the wavelength λ and the carrier density N,

$$\alpha_N = \alpha_{N_0} + \alpha_{N_1} \frac{\lambda - \lambda_\alpha(N)}{[\lambda_z(N) - \lambda]^{1/2}}. \tag{2.41}$$

The bandgap wavelength λ_z is given in Eq. (2.21). As discussed in [85], the quantity α_N becomes infinite at λ_z, because the change of the gain (with respect to N) becomes zero due to strict momentum conservation under the assumption of parabolic bands. The parameter λ_α depends on the bandgap wavelength. It needs to be adapted to the bandgap wavelength of the sample under test. λ_α is parameterized around the transparency point N_0,

$$\lambda_\alpha = \lambda_{\alpha 0} + c_\alpha(N - N_0). \tag{2.42}$$

The quantity c_α defines the dependency of λ_α from the carrier density N. It can be shown that Eq. (2.41) corresponds to a first order Taylor series expansion of the numerator in Eq. (2)

from [85] with physical parameters derived from experiment [39]. The α_N modeled is shown in Fig. 2.6.

Applying Eq. (2.20), (2.21), (2.23) and (2.24) together with the appropriate α-factors allows us to determine the phase evolution in a wavelength conversion. Following Eq. (2.14) and using Eq. (2.19), the respective equations for the nonlinear phase shifts of the forward and backward traveling waves, $\Delta\varphi_i^+(z,t)$ and $\Delta\varphi_i^-(z,t)$, are:

$$\frac{\partial\Delta\varphi_i^+(z,t)}{\partial z}+\frac{1}{\upsilon_g}\frac{\partial\Delta\varphi_i^+(z,t)}{\partial t}=\frac{1}{2}[\alpha_{N,i}\,\Delta(\Gamma_i\overline{g}_{m,i}-\alpha_{int,i})$$
$$+\alpha_T\Delta(\Gamma_i\sum_{\beta=c,v}g_{T,\beta,i})+\alpha_{TPA}\,\Delta(g_{TPA,i})],\tag{2.43}$$

$$\frac{\partial\Delta\varphi_i^-(z,t)}{\partial z}-\frac{1}{\upsilon_g}\frac{\partial\Delta\varphi_i^-(z,t)}{\partial t}=-\frac{1}{2}[\alpha_{N,i}\,\Delta(\Gamma_i\overline{g}_{m,i}-\alpha_{int,i})$$
$$+\alpha_T\Delta(\Gamma_i\sum_{\beta=c,v}g_{T,\beta,i})+\alpha_{TPA}\,\Delta(g_{TPA,i})],\tag{2.44}$$

where $\overline{g}_{m,i}\equiv\overline{g}_m(\lambda_i)$ and $g_{T,\beta,i}\equiv g_{T,\beta}(\lambda_i)$. Note that we implicitly assume in Eq. (2.43) and (2.44) that the nonlinear phase shift in the stationary state is zero, e.g., before the control signal arrives. The difference terms $\Delta(X)$ in Eq. (2.43) and (2.44) stand for the respective change of the linear modal gain, the CH gain compression and the TPA gain compression. They are calculated from their instantaneous values with respect to their values in the stationary state.

Equations for spontaneous emission photon density

Besides the input signals present in the SOA, there are noise signals which are also amplified along the SOA, namely the noise due to amplified spontaneous emission (ASE). The ASE signals also obey the same traveling-wave equation. When reflecting facets are present, the spontaneously emitted noise will visualize the presence of longitudinal modes in the spectral range of interest, i.e. the linewidth of SOA. For this reason, it may be assumed that noise photons only exist at discrete frequencies, which have an equidistant longitudinal mode spacing Δf in Eq. (2.25). All the longitudinal ASE modes available in the spectral range of interest, i.e., the linewidth of SOA, are indexed with k. They are also decomposed into forward- and backward-propagating waves with respect to the z direction, and can be expressed similarly as Eq. (2.19). However, it is sufficient to describe the spontaneous emission in terms of power. The reason is because the phases between adjacent longitudinal ASE modes are random. Similar to Eq. (2.35) and (2.36), the photon densities of the forward ASE waves $s_k^{ASE,+}(z,t)$ and the backward ASE waves $s_k^{ASE,-}(z,t)$ can be described by the multi-mode rate equations:

$$\frac{\partial s_k^{ASE,+}(z,t)}{\partial z}+\frac{1}{v_g}\frac{\partial s_k^{ASE,+}(z,t)}{\partial t}=(\Gamma_k g_{m,k}-\alpha_{int,k})s_k^{ASE,+}(z,t)+R_{sp,k}\,,\tag{2.45}$$

$$\frac{\partial s_k^{ASE,-}(z,t)}{\partial z}-\frac{1}{v_g}\frac{\partial s_k^{ASE,-}(z,t)}{\partial t}=-(\Gamma_k g_{m,k}-\alpha_{int,k})s_k^{ASE,-}(z,t)-R_{sp,k}\,.\tag{2.46}$$

The term $R_{\mathrm{sp},k}$ in (2.45) and (2.46) is the spontaneous radiative recombination. This term is discussed in the following.

Radiative Spontaneous Emission

The radiative spontaneous emission $R_{\mathrm{sp},k}$ in Eq. (2.45) and (2.46) describes the coupling of all the spontaneously emitted photons within the frequency spacing Δf, see Eq. (2.25), into the k-th longitudinal ASE mode. We assume that the spontaneously emitted photons spread uniformly over this frequency spacing Δf. The spontaneous term $R_{\mathrm{sp},k}$ can be calculated by using a quantum-mechanical expression [9] and [84], with the help of the inversion factor n_{sp} and the gain coefficient $\overline{g}_{m,k}$,

$$R_{\mathrm{sp},k} = \frac{\Gamma^2\, n_{\mathrm{sp}}\, \overline{g}_{m,k}\,\Delta f}{v_g A} = \frac{\Gamma^2\, n_{\mathrm{sp}}\, \overline{g}_{m,k}}{2V_{\mathrm{act}}}, \tag{2.47}$$

where $V_{\mathrm{act}} = AL$ is the volume of the active region.

2.2.4 Multi-Section Numerical Algorithm to Solve Rate Equations

As differential equations (2.33), (2.35), (2.36), (2.45) and (2.46) are coupled and cannot be solved analytically, a numerical algorithm should be implemented under consideration of the boundary conditions imposed at the amplifier facets.

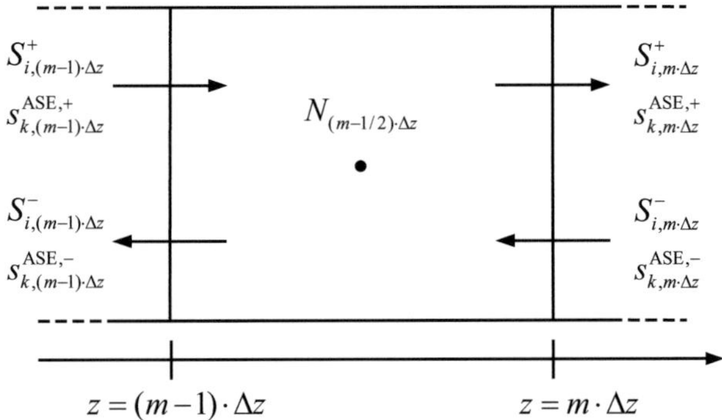

Fig. 2.7. m-th section of the SOA model. The photon densities of the input signals and the spontaneous emission signals are evaluated at the section boundaries, while the carrier density is calculated at the center of the section.

The numerical algorithm is built in the following way: Along the propagation direction, the SOA is divided into a number of sections with a length of $\Delta z = \Delta t \cdot v_g$, where Δt is the time step used to solve the rate equation for the carrier density. The m-th section is illustrated in Fig. 2.7. It can be seen that, for each time step n, the photon densities of the input signals and the spontaneous emission signals are evaluated at the boundaries of each section, $m \cdot \Delta z$

and $(m-1)\cdot\Delta z$, while the carrier density is estimated at the center of the section $(m-1/2)\cdot\Delta z$. For the first and the last section, the photon densities of the input signals and the spontaneous emission signals are subject to the boundary conditions

$$S_{i,0}^{+} = (1-R_1)\,S_i^{+} + R_1\,S_{i,0}^{-}\,, \quad s_{k,0}^{\mathrm{ASE},+} = R_1\,S_{k,0}^{\mathrm{ASE},-}\,, \tag{2.48}$$

$$S_{i,\mathrm{end}}^{-,i} = (1-R_2)\,S_i^{-} + R_2\,S_{i,\mathrm{end}}^{+}\,, \quad s_{k,\mathrm{end}}^{\mathrm{ASE},-} = R_2\,s_{k,\mathrm{end}}^{\mathrm{ASE},+}\,, \tag{2.49}$$

where R_1 and R_2 are power reflectivities at the left and right facets. In this work, they take a value of 10^{-4}. S_i^{+} and S_i^{-} are the forward and backward input photon densities to the left and right facets.

Because the differential equations for carrier density and gain compression terms, Eq. (2.23) and (2.33), are stiff, a predictor-corrector method is implemented to obtain an implicit Euler's solution with a residual error in the order of $O(\Delta z)$. Details of this predictor-corrector method are given in Appendix C.

The calculation of a numerical XGM experiment shown in Fig. 1.2 is performed as follows. First, a preliminary calculation was carried out without the control signal to get a stationary spatial profile, which includes the longitudinal distribution of the carrier density, the photon densities of longitudinal ASE modes and the photon density of the input probe signal along the SOA. After the preliminary calculation, the input control signal is then added.

2.3 Simulation Results of Basic SOA Characteristics

The numerical model presented in section 2.2 is now used to analyze the SOA gain characteristics and the spatial distributions of the carrier density inside the SOA. To probe the gain of the SOA, a single continuous wave (cw) signal is launched in the SOA.

2.3.1 Gain Characteristics

When a single cw signal is present in an SOA, the increase of the signal power increases the carrier depletion, leading to a decrease of the amplifier gain. This limits the achievable gain when the SOA is used in wavelength conversion.

Typical SOA gain (fiber-to-fiber gain) characteristics versus input and output signal power are shown in Fig. 2.8(a) and (b). In the simulation, the input power of the cw signal varies from -30 to 5 dBm, while the cw wavelength is 1550 nm, 1560 nm and 1570 nm. To qualifying gain saturation, we use a parameter input saturation power, which is defined as the input power at which the amplifier gain is 3 dB less than the small signal gain. For instance, read from Fig. 2.8(a), the input saturation power for a cw wavelength of 1550 nm is about -15 dBm while the amplifier gain decreases from ~19 dB to ~16 dB. We can also use a parameter output saturation power, which is defined as the output power at which the amplifier gain is 3 dB less than the small signal gain. For instance, read from Fig. 2.8(b), the output saturation power for a cw wavelength of 1550 nm is about 1 dBm. Simulation parameters are given in Appendix B.

Fig. 2.8(c) shows the gain saturation characteristic versus the bias current. We see that the amplifier gain at a longer wavelength first goes into saturation with respect to the increase of the bias current, e.g. the amplifier gain at 1570 nm in Fig. 2.8(c). The simulation results coincide with the fact that the low (high) levels inside the conduction (valence) band are first filled by the electrical pumping.

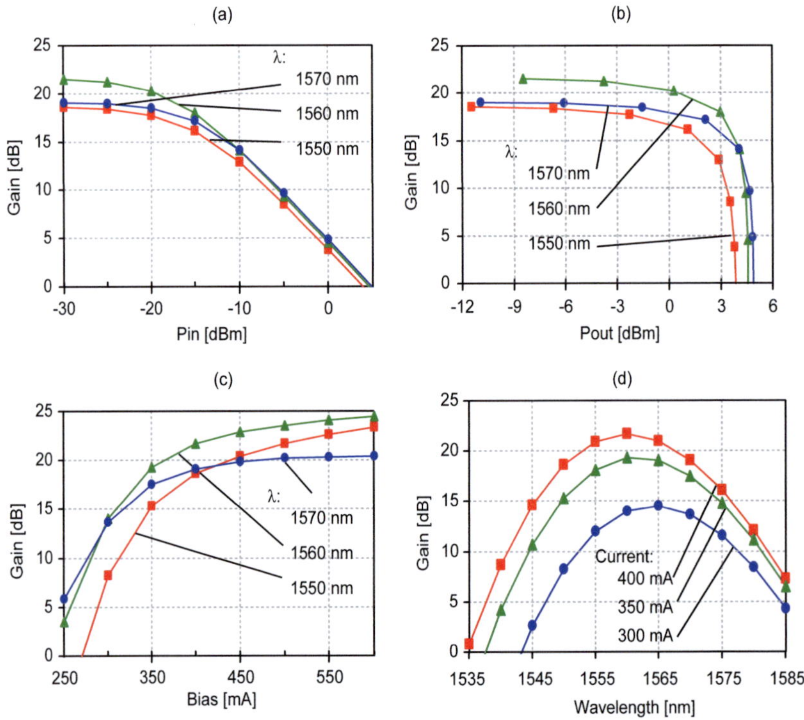

Fig. 2.8. Simulated fiber-to-fiber gain characteristics of an SOA with a length of 2.6 mm. (a) and (b) Characteristics of the gain saturation with respect to the input and output cw power at a wavelength of 1550 nm, 1560 nm and 1570 nm. The bias current is 400 mA. (c) Characteristics of the gain saturation with respect to the bias current, while the input cw power is -35 dBm at a wavelength of 1550 nm, 1560 nm and 1570 nm. (d) Gain spectra with respect to the wavelength, while the bias currents are 400 mA, 350 mA and 300 mA, respectively. Note that the ASE noise is not included in the calculation.

The small signal gain spectra for different bias currents are simulated and shown in Fig. 2.8(d). Increase of the current bias moves the gain peak to a lower wavelength (high frequency) in the spectrum. This result can be understood as follows. At a low carrier density, low energy levels in the bands are first filled. As the carrier density increases along with the increase of the bias current, high energy levels will be filled. As a consequence, the gain peak moves to a high frequency (high energy, lower wavelength). However, as the bias current further increases, the available amplifier gain is clamped due to the gain saturation

2.3.2 Spatial Carrier Distribution inside an SOA

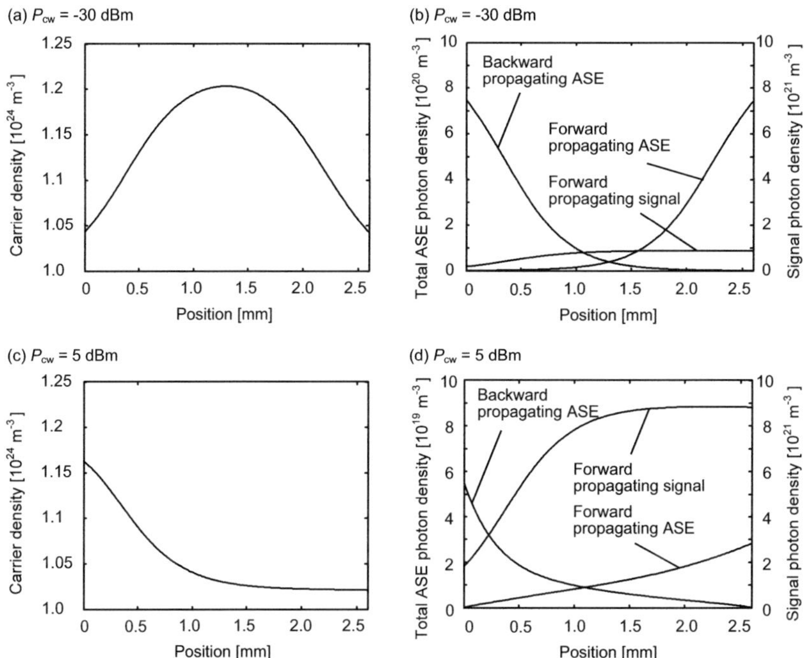

Fig. 2.9. Simulated spatial distribution with respect to position z inside an SOA. (a) and (c) are the spatial carrier density distributions with an input cw signal power of -30 dBm and 5 dBm, respectively. (b) and (d) are the spatial distributions of total forward and backward propagating ASE photon densities with an input cw signal power of -30 dBm and 5 dBm, respectively. (b) and (d) also show the propagation for the input cw photon density along the SOA.

Using the SOA model above, we can also analyze the spatial distributions of the carrier density, ASE photon densities and necessarily the photon densities of input signals inside an SOA. These characteristics are not measurable in the experiment. Fig. 2.9(a) and (b) show these distributions inside an SOA for a low input cw signal power of -30 dBm. At such a lower cw power, the carrier density has a symmetrical spatial distribution showing a peak at the center of the SOA and decreasing toward the input facet (position $= 0$ mm) and output facet (position $= 2.6$ mm), Fig. 2.9(a). This is because the ASE photon densities are highest in regions near the facets as shown in Fig. 2.9(b). The input cw power is amplified along the SOA. At a high input cw signal power of 5 dBm, the carrier density spatial distribution becomes more asymmetrical, Fig. 2.9(c). The peak moves toward the input facet, as the more carriers are depleted along with the amplification of the input signal. The signal photon density also dominates over ASE photon densities, especially of the forward propagating ones, Fig. 2.9(d). Note that, as the facet reflectivities are low (10^{-4}), no ripples can be observed in the carrier density distributions, Fig. 2.9(a) and (c).

Analysis of SOA characteristics is useful in designing an SOA. This also helps us to operate the SOA properly in wavelength conversion. The results given in Fig. 2.8 and Fig. 2.9 show the SOA characteristics, as only one cw signal is present. In the next Chapter we discuss how an SOA responds to a second control signal is launched into SOA.

3 Cross-Gain and Cross-Phase Modulation

The wavelength conversion experiments, investigating the SOA response both to a single Gaussian pulse as well as to bit patterns at 160 Gbit/s, are shown and compared with the simulations. The temporal dynamics of the effective α-factor is also investigated in this Chapter. Unlike a constant α-factor assumed in literature, this effective α-factor during cross-gain modulation (XGM) experiments varies strongly with time. It even takes on negative values for short periods of time. As a consequence, cross-phase modulation (XPM) effects usually lag behind XGM effects by several picoseconds.

3.1 Experimental Scheme

The setup used for the cross-gain modulation (XGM) experiment [82] is shown schematically in Fig. 3.1(a). The SOA under test has a length of 1.6 mm and tilted facets with reflectivities below 10^{-4}. A control signal P_{in}, Gaussian pulse having a full-width at half maximum (FWHM) of 3 ps from a mode-locked laser, is launched into the SOA together with a continuous wave (cw) probe signal P_{cnv}. They are combined in an optical coupler having a 50:50 splitting ratio.

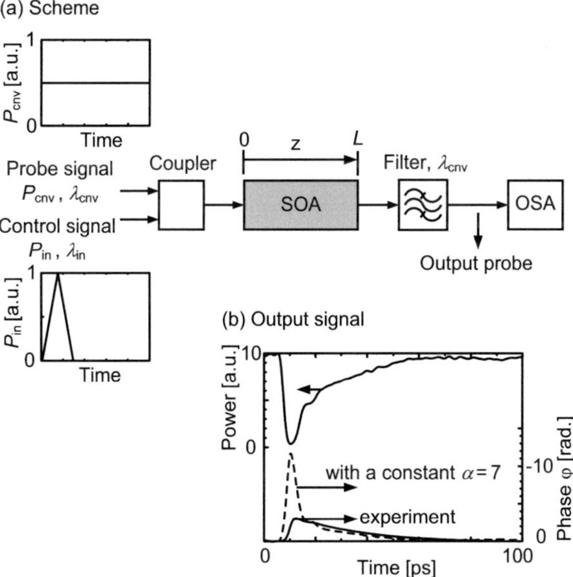

Fig. 3.1. (a) Schematic cross-gain modulation experiment setup. OSA: optical spectrum analyzer. (b) illustrates the power and phase of an output probe signal modulated by an input control signal with a power P_{in} of 5 dBm. In (b) we also plot the hypothetical phase evolution (dashed line), obtained with a device having a constant α-factor of 7. This phase evolution is significantly off from the real phase evolution gotten in an experiment [82].

In the XGM experiment, the cw probe signal experiences also a cross-phase modulation (XPM). At the SOA output, the gain and phase dynamics of the probe signal are determined by using the method of spectrograms [11]. In this method, the modulated probe signal is characterized by measuring its optical spectrum as a function of the optical frequency and relative delay between the control and probe signal [32]. The complete information of the signal, the amplitude and the phase, is then retrieved from this two-dimensional spectrogram [77]. Exemplary output power and the phase shift relaxation are shown on the left and right axis in Fig. 3.1(b).

It is noticeable that the measured phase evolution is far off from that obtained from a hypothetical constant α-factor of 7, dashed line in Fig. 3.1(b). While the phase evolution derived from a constant α-factor does show a fast recovery, the true phase evolution in the experiment shows hardly any fast recovery effects. Therefore, the α-factor during the XGM experiment is not constant. Next, we will analyze the gain and phase dynamics with the help of the simulation, and assess the dynamic behavior of the α-factor.

3.2 Gain, Phase and Alpha-Factor Dynamics

Published by J. Wang, A. Maitra, C. G. Poulton, W. Freude, and J. Leuthold, "Temporal dynamics of the alpha factor in semiconductor optical amplifiers," *IEEE J. Lightw. Technol.*, vol. 25, pp. 891-900, Mar. 2007.

3.2.1 Experimental Results for a Single Input Pulse

We first investigate the SOA gain and phase response to a single input pulse. Gaussian pulses (control signal P_{in}) of $+5$, 0 and -5dBm at a central wavelength of 1558 nm are launched into the SOA together with a cw probe signal P_{cnv} at 1550 nm. All power levels are measured in the fiber before the coupler. Measured power and phase evolutions of the cw signal at the output of the SOA are plotted on the left and right axes in Fig. 3.2(a) for a cw power of 10 dBm, and in Fig. 3.2(b) for a cw power of -2.4 dBm.

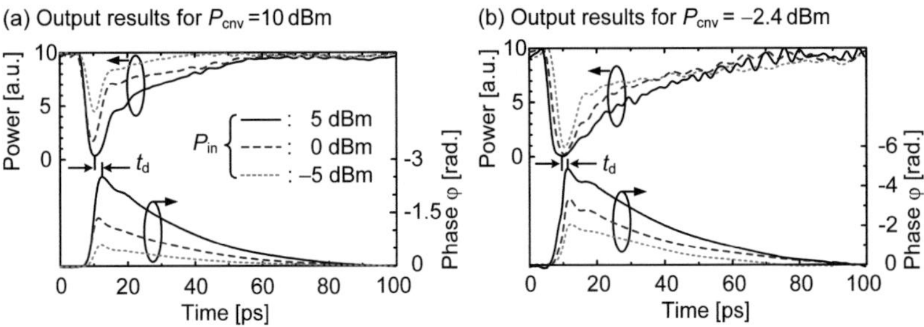

Fig. 3.2. Output power (left axis) and phase (right axis) in cross-gain modulation experiment for control signal P_{in} of $+5$, 0 and -5 dBm, while the cw-probe powers P_{cnv} were 10 dBm in (a) and -2.4 dBm in (b). (a) and (b) also show that the phase change lags the power transmission by a time t_d.

It can be seen in Fig. 3.2, that after launching the Gaussian pulse into the SOA the cw signal is first strongly suppressed and then relaxes back to the initial level. For pump signals with moderate power, one notices first a fast gain recovery followed by a slower gain recovery. The fast effect is due to ultra-fast processes such as spectral hole burning (SHB), carrier heating (CH) and two photon absorption (TPA). The slower relaxation time is due to the carrier recovery in the conduction and valence band. The ripples on the slow-recovery part of the curve are caused by residual reflectivities on the SOA's facets.

Different to the fast gain recovery, the measured phase evolution shows hardly any fast phase recovery effects. While fast nonlinear effects such as CH, SHB, and TPA contribute significantly to the intensity modulation, their contributions to the phase modulation are hardly seen in Fig. 3.2. Thereafter, slow carrier recovery dominates in the phase evolution. Next we relate the gain and phase dynamics via an α-factor.

To assess the dynamic behavior of the α-factor during a XGM experiment, we use the definition of an effective α-factor of a device. It relates the change of the single pass gain $G_s(t)$ (Eq. (2.30)) and the phase shift $\Delta\varphi(t)$ (Eq. (2.13)) of a signal after an SOA. For an SOA with a length of L, the single pass gain $G_s(t)$ is defined in Eq. (2.30). The phase shift $\Delta\varphi(t)$ defined in Eq. (2.13) is the difference between the instantaneous phase $\varphi(t)$ and the phase φ_{cnv} in the stationary state. The phase shift $\Delta\varphi(t)$ is related to the change of refractive index Δn_r, which includes all the contributions due to band filling, CH, SHB and TPA. As discussed in Eq. (2.13) and (2.15), the change of the refractive index Δn_r and the change of the gain are implicitly dependent on the input power. From Eq. (2.43) and (2.37)-(2.40), the phase shift $\Delta\varphi(t)$ at the output can be calculated as

$$\Delta\varphi(t) = \varphi(t) - \varphi_{\mathrm{cnv}} = -k_0 \int_0^L \Delta n_r \left(N(t,z) \right) \mathrm{d}z . \tag{3.1}$$

We can assume that the phase φ_{cnv} in the stationary state takes a value of 0. This leads us to an effective α-factor for the device,

$$\alpha_{\mathrm{eff}}(t) = -\frac{4\pi}{\lambda} \frac{\partial n_r / \partial t}{\partial g / \partial t} = -\frac{4\pi}{\lambda} \frac{\dfrac{1}{L}\int_0^L \Delta n_r \left(N(t,z) \right) \mathrm{d}z}{\dfrac{1}{L}\int_0^L \Delta g \left(N(t,z) \right) \mathrm{d}z} = \frac{2\Delta\varphi(t)}{\Delta[\ln G_s(t)]} . \tag{3.2}$$

Eq. (3.2) actually relates the spatially averaged nonlinear refractive index change along the SOA to the spatially averaged net modal gain change. For small variations of the carrier density N, this effective α-factor will be constant in time.

Applying Eq. (3.2) on the results in Fig. 3.2, we obtain the evolution of the effective α-factor and display them in Fig. 3.3 for a time window of 50 ps. In the calculation, we have assumed that the phase in the stationary state is 0. Thus, the phase shift is then the temporal variation part of the signal phase. The α-factors for cw powers of 10 dBm and -2.4 dBm, respectively, are compared in Fig. 3.3(a) and (b). We see that the effective α-factor varies strongly. It can be seen that the temporal α-factor evolution is almost identical – independent of the cw or control signal power applied. This can be understood from the fact that the gain and phase relaxation are mainly governed by the various carrier relaxation times. As these relaxation times are material constants, the evolution of the α-factor does not change too much either.

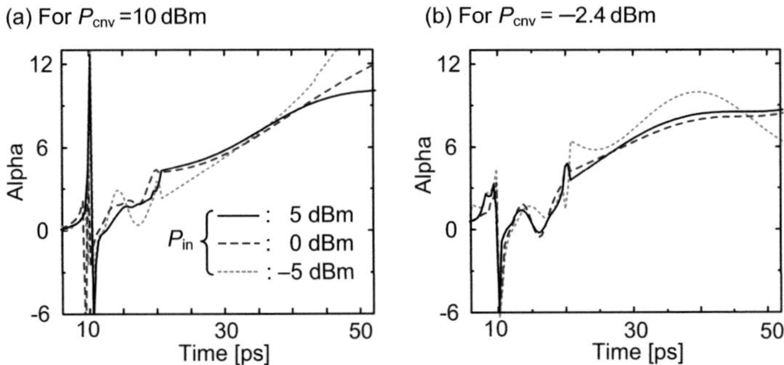

Fig. 3.3. Evolution of the effective α-factor in the cross-gain modulation experiment for control signal P_{in} of +5, 0 and -5 dBm, while the cw-probe powers P_{cnv} were 10 dBm in (a) and -2.4 dBm in (b). Results are obtained by applying Eq. (3.2) to the data from Fig. 3.2.

The origin of the strong time dependence of the effective α-factor lies in various physical effects, with different time constants. The first few picoseconds during a XGM experiment are governed by fast nonlinear effects such as CH, SHB, and TPA. While they contribute significantly to the intensity modulation, they hardly contribute to the phase modulation. In fact, the α-factor arising from these effects is usually small and therefore the phase of the probe signal hardly changes. During the XGM experiment, ultrafast carrier depletion and bandgap shrinkage effects take place as well. Subsequently, the carriers recover. While the carrier depletion and the bandgap shrinkage are fast, carrier recovery takes more time and dominates thereafter. The α-factor related to band-filling (i.e. carrier depletion and carrier recovery) is not constant as usually assumed. It strongly depends on the carrier density itself. As a matter of fact, since the α-factor is quite small for low carrier densities, it contributes little to phase modulation during the first few picoseconds.

The experiment also shows a time delay t_d between the minimum of the transmission and the peak of the phase shift of the modulated cw signal. This delay has been predicted in [57] and been observed previously in [32]. Details of this time delay will be discussed in next section with the help of a simulation.

Finally we observe oscillations in the intensity relaxation curve approximately 6 ps after the main Gaussian pulse. These oscillations have their origin in a ghost pulse that follows the main pulse from the mode-locked laser used in the experiment. As a consequence we also have oscillations in the α-factor

3.2.2 Simulation Results and Comparison

Applying the SOA model from Chapter 2, the XGM experiment was simulated. The simulated output power and phase evolution are compared with the experimental results and shown in Fig. 3.4. Fig. 3.4(a) and (b) are the results in a time window of 100 ps for a control signal of 0 dBm and 5 dBm, respectively, while the cw signal power is 10 dBm. Note that we plot the phase shift instead of the optical phase in Fig. 3.4, as we assume that the phase in the stationary state is 0. We find good agreement between simulated (dashed lines) and measured results (solid lines).

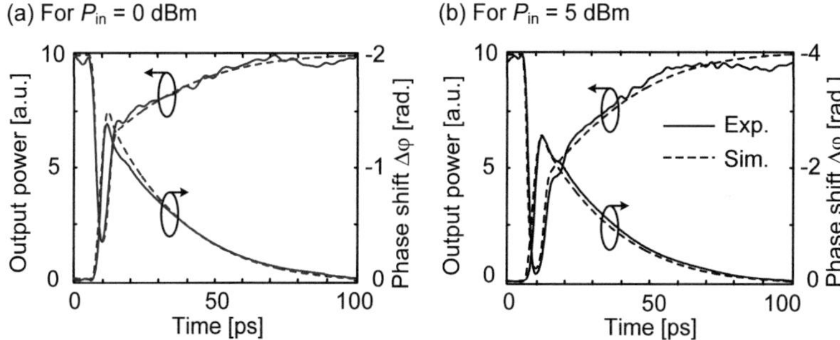

Fig. 3.4. Simulated output power (left axis) and phase relaxation (right axis), dashed lines, are compared with experiment results, solid lines, for (a) P_{in} = 0 dBm and (b) P_{in} = 5 dBm, while the cw-probe power P_{cnv} = 10 dBm. Simulated and experiment results are in good agreement.

A phase variation over the time means a variation of the instantaneous frequency, i.e. a frequency chirp. The frequency chirp is defined as

$$\Delta f = \frac{1}{2\pi}\frac{\partial\varphi}{\partial t} = \frac{1}{2\pi}\frac{\partial(\Delta\varphi)}{\partial t}. \tag{3.3}$$

Note that the phase φ_{cnv} in Eq. (3.1) takes a value of 0. Thus the optical phase φ is replaced by the phase shift $\Delta\varphi$ in Eq. (3.3). Applying Eq. (3.3) on the simulated phase evolutions in Fig. 3.4(a) and (b), we get the frequency chirps and display them in Fig. 3.5.

Fig. 3.5 shows that there are two regimes of the frequency chirp. As the phase decreases the output signal experiences a red chirp (decrease of the frequency). At the maximum phase shift, the frequency chirp is 0. In the phase recovery regime, there is a blue chirp (increase of the frequency). Comparing Fig. 3.5(a) and (b), we see that a stronger input power induces a larger frequency chirp. In both input powers, the maximum red chirp is larger than the maximum blue chirp. The reason lies on the different carrier concentrations at these two points. At the maximum blue chirp, the carrier concentration has not yet recovered and the material gain is smaller than that at the maximum red chirp. The material gain in turn determines the carrier depletion (recovery) rate. With an amplitude-phase coupling via a linewidth enhancement factor, the maximum red chirp is then larger than the maximum blue chirp.

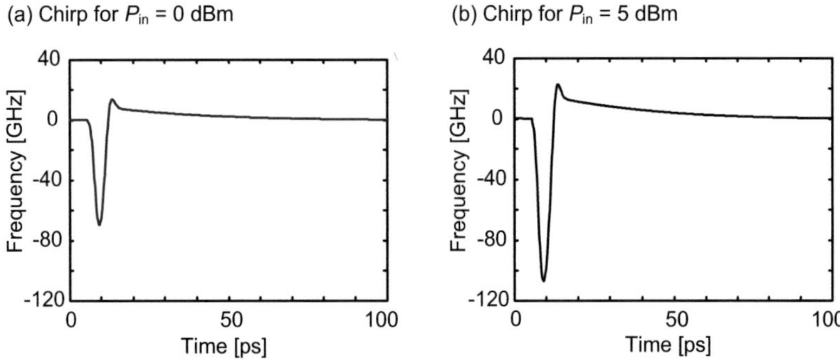

Fig. 3.5. Calculated frequency chirp in cross-gain modulation. Results are obtained by applying Eq. (3.3) to Fig. 3.4. The control signal powers P_{in} were (a) 0 dBm and (b) 5 dBm, while the cw-probe power P_{cnv} = 10 dBm.

3.2.3 Time Delay between Cross-Gain and Cross-Phase Modulation

It is worth discussing the time delay observed in both experiment and simulation. This leads to a better understanding of the temporal evolution of the effective α-factor. We look at a magnified time window of 18 ps from Fig. 3.4(b) and Fig. 3.3. The evolution of output power and phase are shown in Fig. 3.6 and the evolution of the effective α-factor is shown in Fig. 3.7. In order to simplify the discussion we have split the time interval into 4 windows.

Our discussion starts with window A, when the control signal is launched into the SOA. In this window the amplitude of the cw signal is suppressed because of carrier depletion in the bands but also because of the fast compression effects such as SHB and CH. The phase on the other hand is basically effected by the carrier depletion – and only to a small extent by the gain compression effects. As a result we obtain a positive and increasing effective α-factor due to strong gain depletion with simultaneous weak phase shift. The signal is red-shifted.

Window B: As the gain saturation becomes strong, carrier depletion slows down. At one point, the carrier depletion will be exactly compensated for by the much faster reallocation of hot carriers due to SHB and CH. As a result, the gain change becomes zero – while the red-chirped phase effect still continues because of a slower but ongoing carrier depletion effect. In this time window, the α-factor increases, and approaches infinity at precisely that point when the change in gain vanishes.

Window C: As the gain recovers, the α-factor comes back from negative infinity. When the phase shift peaks around 2 ps after the transmission minimum (for a given pulse width (FWHM) of 3 ps), there is no further phase shift for a short moment, i.e. a zero frequency chirp Δf. In this time window, the α-factor sweeps from negative infinity to zero.

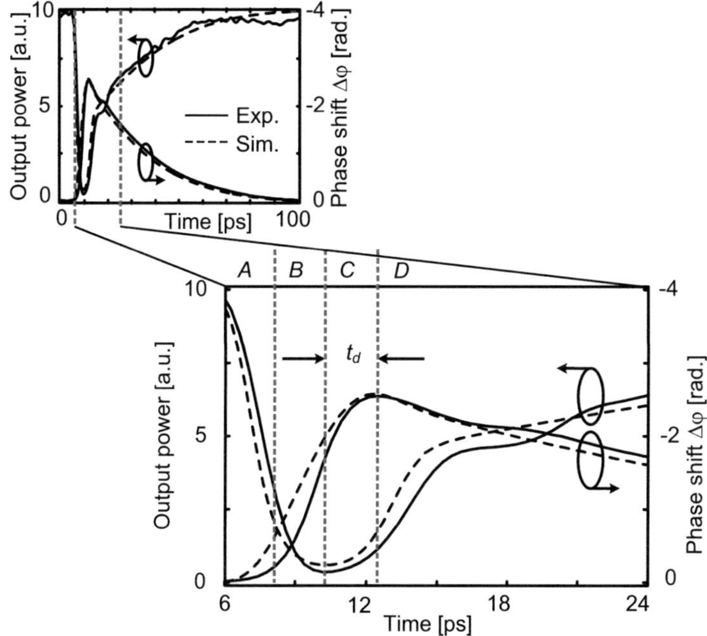

Fig. 3.6 Snapshot of probe signal dynamics in a time window of 18 ps at the SOA output for P_{in} = 5 dBm and P_{cnv} = 10 dBm. Calculated (dashed) and experiment results (solid) both show that the phase change lags the power transmission by a time t_d. A, B, C and D are respective time windows.

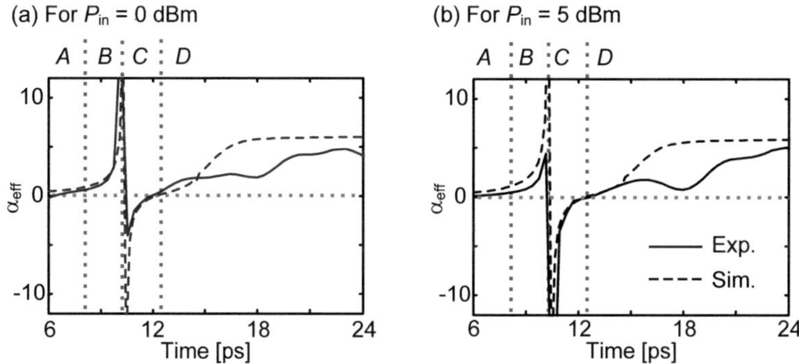

Fig. 3.7. Simulation results of the effective factor evolution in the cross-gain modulation experiment. Snapshots of calculated (dashed) and experiment results (solid) in a time window of 18 ps are shown for (a) P_{in} = 0 dBm and (b) P_{in} = 5 dBm. A, B, C and D are respective time windows.

Window D: After the Gaussian pulse has left the SOA, the band-filling dominates. As the electrical pumping is refilling the bands, the α-factor increases as discussed by Eq. (2.41). The fast effects hardly contribute any longer. The oscillation in the experimental curve of the α-factor around the position of 18 ps, seen in Fig. 3.7, has its origin in a ghost pulse that follows the main control pulse of our mode locked laser.

The contributions of each effect to the total phase shift are plotted in Fig. 3.8. In this figure, $\delta\varphi$ is the phase difference at a particular time between the output cw wave and the input cw wave through an SOA. Fig. 3.8 shows the contributions from the band-filling effect (solid line), the carrier heating (dashed lines) and the two photon absorption (dotted lines.). The calculations confirm that the effects from the ultrafast nonlinearities, CH and TPA, onto the phase are quite small.

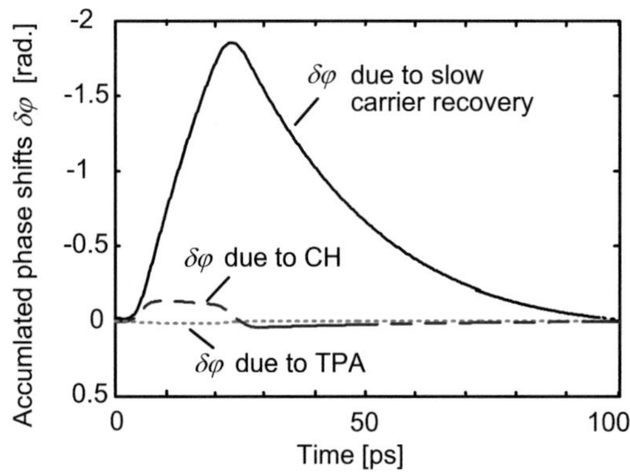

Fig. 3.8. Proportionate contributions of the various effects to the phase shift passed on to the cw probe signal during an XGM experiment. The fast CH and TPA effects contribute only little to the total phase shift, while the band-filling effect dominates. In this simulation, $P_{in} = 5$ dBm and $P_{cnv} = 10$ dBm.

Fig. 3.6 also shows a time delay t_d between the maximum of the XGM and the maximum of the XPM. This delay is due to the aforementioned discrepancies of the ultrafast effects contributing considerably to the gain changes but only little to the phase changes. Fig. 3.9 shows the absolute time delay between XGM and XPM (in picoseconds on the left axis) and the relative delay with respect to the FWHM of the input signal for different FWHMs of input pulses. The input powers are 5 dBm (solid lines) and 0 dBm (dashed lines), respectively, while the cw-power is 10 dBm. While absolute delay becomes smaller with shorter pulses, the relative delay increases dramatically. For instance, in an experiment with input-pulses of a FWHM of 3 ps the XPM will peak 2 ps after the signal has passed the SOA. The value of time delay does not change much for different input powers.

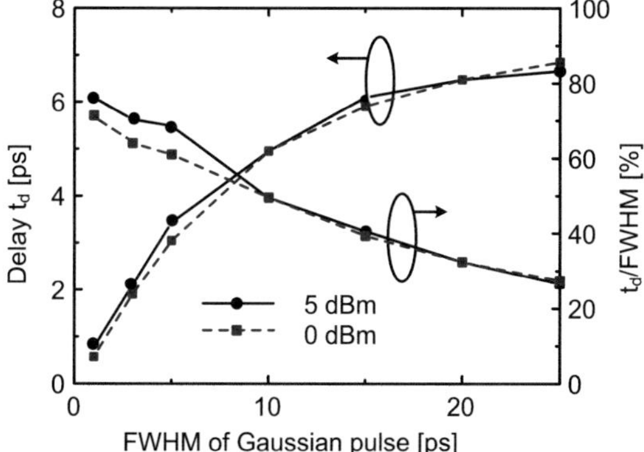

Fig. 3.9. Left axis: Simulated time delay t_d between the maximum gain suppression (maximum XGM) and the maximum phase shift (maximum XPM) as a function of the control signal FWHM. Right axis: Ratio between t_d and the FWHM of respective control signal pulse. The control signal powers P_{in} were 5 dBm (solid lines with round marker) and 0 dBm (dashed lines with square marker), while the cw-probe power $P_{cnv} = 10$ dBm.

The delay is particularly worrisome with regard to the usage of the XPM effect for ultrafast signal processing. At higher speed, the time delay of the phase in XPM-based devices will not only be of minor importance. It can also translate into a timing jitter. This might require a rethinking of simulations that have been performed in the past with regards to the usage of purely XPM based devices for high-speed all-optical communications.

3.3 Characteristics of Cross-Gain Modulation

The XGM experiment conditions have a great influence on the gain and phase dynamics. The gain and phase dynamics are practically characterized by a carrier recovery time. A large carrier recovery time of an SOA means that the SOA is slow. In a XGM, a very slow SOA shows a pattern effect, which is a serious problem in high-speed all-optical communications. In the following, we describe the relationship of the carrier lifetime to the experiment conditions, e.g. the bias current and the SOA length. We will also compare the carrier recovery in a configuration, in which a cw-probe signal co-propagates and a second cw-probe signal counter-propagates with respect to the control signal. Finally, the pattern effect will be discussed.

3.3.1 Carrier Recovery Time

Influences of the SOA bias and input power

The influences of the bias current and the probe power are investigated here. In an SOA-based experiment, the bias current is needed for two major purposes. It first electrically pumps an SOA, offering power amplification to the present input signal. After the carrier depletion due to input signal, the bias current refills the bands. The refilling process is characterized by a carrier recovery time. The recovery time discussed here is the 10% to 90% phase recovery times of the SOAs and not the gain recovery times. The reason that we choose the phase recovery times for reference is that the gain dynamics usually comprises both the ultrafast dynamics associated with SHB and CH as well as the slower band-filling effect – whereas the phase recovery mostly is influenced by the band-filling dynamics.

XGM simulations have been run under different bias currents and different cw-probe powers. In the simulation, the SOA length is 2.6 mm. The respective carrier recovery times are retrieved and depicted as a function of the cw-probe powers for three different bias currents in Fig. 3.10. The relation between the bias current and the recovery time is clearly demonstrated in Fig. 3.10. It can be seen that an increase of the bias current, i.e. an increase of carrier density in the active region, leads to a decrease of the recovery time. Also, a decrease of the recovery time is observed with an increase of the cw-probe power, i.e. an increase of the photon number inside the active region. Both methods increase the carrier-photon interaction, thus the speed of the SOA. However, with a high bias current e.g. 500 mA, the SOA is saturated even at low cw-probe power. Thus, as shown in Fig. 3.10, the recovery time versus the cw-probe power is almost a flat line.

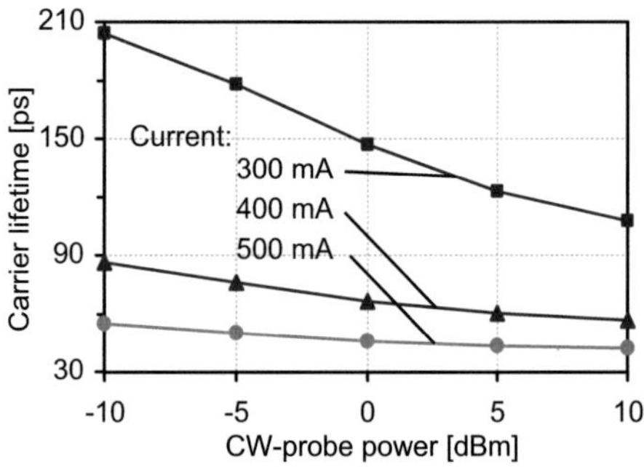

Fig. 3.10. Carrier lifetimes as a function of input cw-probe powers, launched in an SOA with a length of 2.6 mm. The bias currents for the SOA are 300 mA (with square markers), 400 mA (with triangle markers) and 500 mA (with round markers), respectively.

Influence of the SOA length

The influence of the SOA length on the recovery time is of a practical interest. One needs to know how long an SOA should be before it is fabricated. Simulations were run for SOAs having a length of 1.6 mm and 2.6 mm. Other parameters are found in Appendix B. Their recovery times are compared in Fig. 3.11(a) and (b) for applying a same current density of 10.26 kA/cm or a same bias current of 400 mA. For a same current density applied, a shorter SOA exhibits a smaller amplifier gain for a same input cw power. This leads to a less photon number inside the shorter SOA, and thus a larger recovery time. The similar phenomenon has been observed in an experiment [16] and in a simulation [15]. In Fig. 3.11(a), almost a two times large recovery time is observed in a shorter SOA having a length of 1.6 mm, compared to a longer SOA of a length of 2.6 mm. For a same bias current, the current density becomes larger in the 1.6 mm long SOA. The amplifier gain becomes comparable large to that of 2.6 mm long SOA. In Fig. 3.11(b), these two SOAs exhibit almost same recovery times.

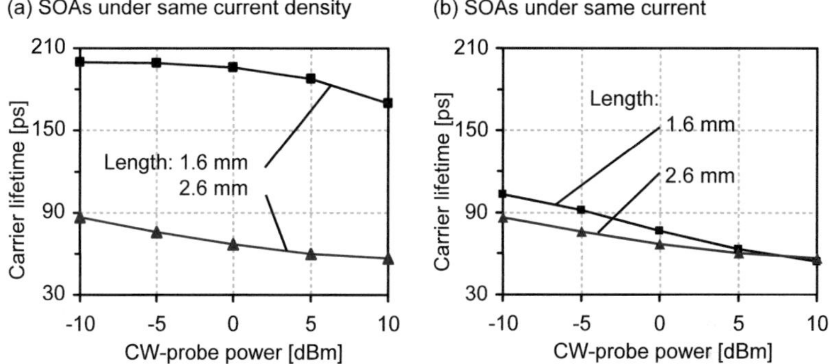

Fig. 3.11. Carrier lifetimes as a function of input cw-probe powers, launched in SOAs with a length of 1.6 mm (with square marker) and 2.6 mm (with triangle marker). Results are compared for (a) a same current density of 10.26 kA/cm and (b) a same current of 400 mA.

3.3.2 Comparison between Co- and Counter-Propagating Signals

In an XGM experiment, the probe signal can either co- or counter-propagate with the control signal. However, the carrier dynamics in these two configurations are different. A scheme having both co- and counter-propagating probe signals, with respect to a control signal at a wavelength λ_{in}, is shown in Fig. 3.12. Both probe signals at a wavelength λ_{cnv} are cw signals. Practically, circulators can be used to separate the output probe signals.

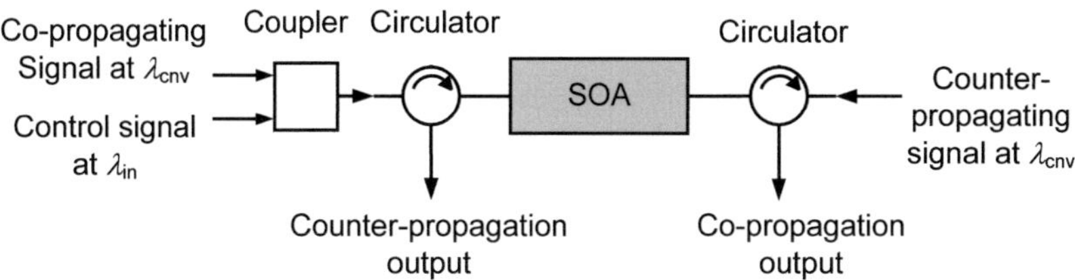

Fig. 3.12. Schematic cross-gain modulation experiment, where one probe signal co-propagates with a control signal (from left to right) through the SOA and another probe signal counter-propagates (from right to left) with the control signal.

The output XGM signals are compared in Fig. 3.13(a) for a 2.6 mm long SOA and in Fig. 3.13(b) for a 0.6 mm long SOA. The control signal is a Gaussian pulse with a FWHM of 25/3 ps. For a longer SOA in Fig. 3.13(a), the output power and phase shift of the co- and counter-propagating signals are asymmetric. Namely, the gain saturation process for a counter-propagating signal takes place earlier and affects longer that that for a co-propagating signal. However, the asymmetric gain and phase responses are not obvious in a shorter SOA, Fig. 3.13(b).

The asymmetric gain saturation and phase shift of two counter-propagating probe signals must dependent on the length of the SOA. As discussed in [31], the gain saturation and phase shift for the co-propagating signal is proportional to the integral over the input control signal. However, also discussed in [31], the gain saturation and phase shift for the counter-propagating signal are not only proportional to the integral over the input control signal, but also proportional to the integral over the SOA length. In other words, the gain saturation for the counter-propagating signal is averaged over the input control signal and the SOA length. As a consequence, the gain saturation process for counter-propagating signal affects over a larger time span. Also, after the control signal passes, the carrier recovery takes place. Since the carrier refilling is mainly material and bias dependent, the recovery process does not exhibits dependence on the signal propagation direction, Fig. 3.13(a) and (b).

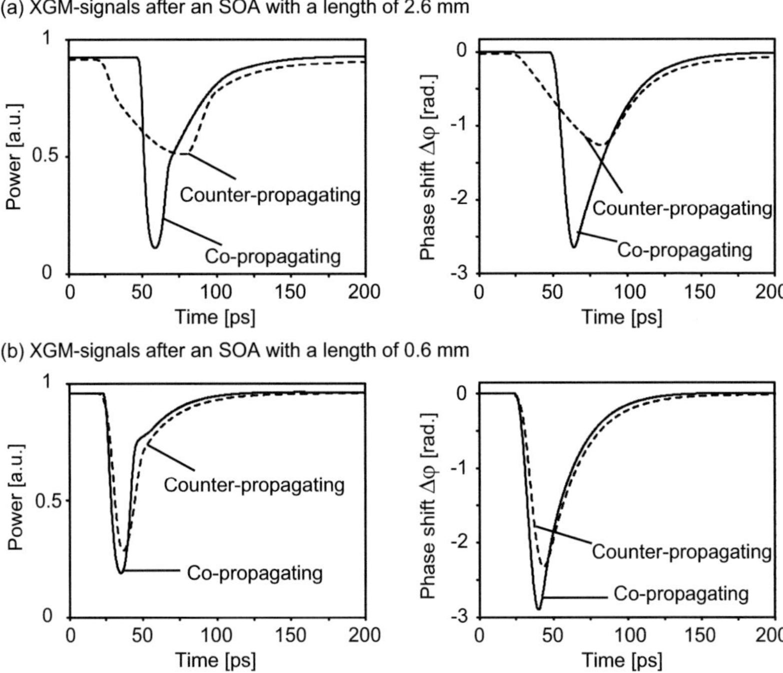

Fig. 3.13. Comparisons between co-propagating (solid line) and counter-propagating (dashed line) XGM signals after an SOA with a length of (a) 2.6 mm or (b) 0.6 mm, where the left and right columns are the output power and phase shift, respectively.

3.3.3 Pattern Effect

Pattern effect in gain and phase dynamics

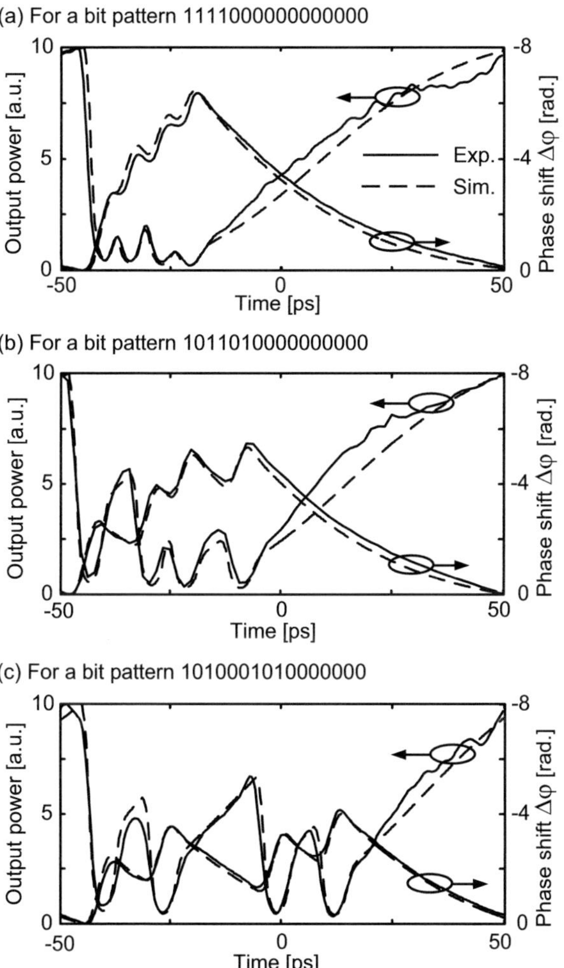

Fig. 3.14. Results of cross-gain modulation experiment with a bit pattern (a) 1111000000000000, (b) 1011010000000000, and (c) 1010001010000000. Output power (left axis) and phase relaxation (right axis) are compared between experiment (solid lines) and simulation (dashed lines).

As subsequent control pulses are launched into a slow SOA, the carrier density is depleted continually. It recovers back to different levels and the amplifier gain also varies for different pulses, depending on the former bit pattern seen by the SOA. This effect is usually called as pattern effect. In our XGM experiment and simulation, 160 Gbit/s control signals with different bit patterns were launched into an SOA with a length of 1.6 mm. The output powers and phase shifts are compared in Fig. 3.14(a) for a bit pattern "1111000000000000", in Fig. 3.14(b) for a bit pattern "1011010000000000", and in Fig. 3.14(c) for a bit pattern "1010001010000000".

We see that the pattern effect will be a serious problem in high-speed signal processing. As shown in Fig. 3.14(a), the phase shift of subsequent bits increases but the relative phase shift between two subsequent bits is different. Sometimes, the pattern effect can be so strong that one bit "0" can not be recognized from bit "1"s, e.g. bit "0" in the bit pattern "1101" in Fig. 3.14(b).

Reducing pattern effect in an SOA can make SOA-based all-optical processing schemes working at high speed. Throughout this work, rather than decreasing the SOA recovery time by proper material design [92] or choice of new fast materials [73], the main attention will be paid to mitigate pattern effect by optimum operation conditions and careful configuration design.

Pattern effect in XPM-induced chirp

Now we investigate the pattern effects in the XPM-induced red and blue chirp of the inverted signal. Simulated frequency chirp for an input control signal with a bit pattern "01111" are given in Fig. 3.15, left y-axis. As a reference, the power gain under this bit pattern is also plotted on the right y-axis in Fig. 3.15. The gain decreases with the leading edge of each input pulse, accompanied with a red chirp, and recovers with the falling edge of each pulse, accompanied with a blue chirp. Further, as the SOA gain gradually saturates for each subsequent bit, seen in Fig. 3.15, also the phase shift saturates. The variation of the phase shift per time increment, i.e. the frequency chirp defined in Eq. (3.3), also undergoes the pattern effect. As we will discuss below, the red and blue chirps have complementary pattern dependences.

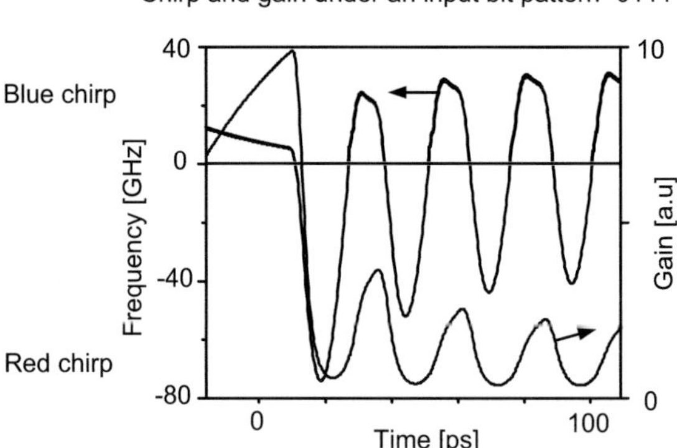

Fig. 3.15. Calculated frequency chirp and power gain, left and right y-axes, in cross-gain modulation for an input bit pattern "01111".

We start for the red chirp. For the first "1" bit in the pulse train, a strong carrier depletion and a large gain reduction take place. This gives a large red chirp on the leading edge, Fig. 3.15. For subsequent "1" pulses, the SOA does not fully recover. The gain of these subsequent "1" pulses is smaller than for the first "1" bit, seen in Fig. 3.15. Therefore, from Eq. (2.33), the carrier depletion rate and also the gain reduction rate decrease. As the gain reduction rate decreases, the induced red chirp decreases as well, low part of Fig. 3.15.

The pattern dependence of the blue chirp in the converted signal can be understood as follows. The gain recovers with the trailing edge of the first "1" pulse, creating a blue chirp. With subsequent bits launched into the SOA, the SOA is further saturated and the amplifier gain decreases with following "1" bits. In turn, the carrier depletion due to stimulated emission is now weaker while the current injection is unchanged. As a consequence, the carrier recovery rate is faster for subsequent pulses. This leads to a stronger blue chirp as depicted in the upper part of Fig. 3.15.

In Chapter 4, we will show how to utilize the pattern effect of the red and blue chirp in the wavelength conversion process. It is realized via the frequency-amplitude conversion at the slope of the optical filter.

4 Pattern Effect Mitigation Techniques in Wavelength Converters Assisted by an Optical Filter

The slow gain recovery in semiconductor optical amplifier (SOA) induces unwanted pattern effects in the converted signal and limits its implementation at high speed. The most practical approach to overcome pattern effects is to decrease the SOA recovery time by proper design [92], optimum operation conditions [16], an additional assisting light [52], and choice of new fast materials [73]. Other approaches to mitigate the pattern effects are cascading several SOAs [5] and [53] or by using SOAs in a differential interferometer arrangement. Among them, the differential Mach-Zehnder interferometer (MZI) [75], the differential Sagnac loop [12], the ultrafast nonlinear interferometer (UNI) [25] and the delay interferometer (DI) configurations [41], [47] and [78], which exploit the XPM effect enable speeds beyond the limit due to the SOA carrier recovery times.

Recently, a new wavelength converter with an SOA followed by a single pulse reformatting optical filter (PROF) has been introduced [45] and [46]. The PROF scheme exploits the fast chirp effects in the converted signal after the SOA. The PROF filter basically represents an optimum filter for the SOA response with the potential for highest speed operation. However, so far it is not clear, if these schemes with optical filters can successfully overcome pattern effects at highest speed. In this Chapter, we will investigate the pattern effect mitigation techniques in these schemes.

4.1 Pattern Effect Mitigation Using a Delay Interferometer Filter

In this section, we will first introduce the configuration of the SOA followed by a delay interferometer (DI). Also, we will give the transfer function of the DI filter. We then discuss the operation principle of the differential DI scheme, where the DI mainly performs the differential operation on the cross-phase modulated (XPM) signal after the SOA. This differential scheme is extremely efficient in speeding up the performance of the SOA, i.e. in mitigating pattern effects.

Similarly to DI scheme for improving XPM there is a technique based on the DI to enhance the performance of XGM. The scheme relies on the fact, that SOA nonlinearities show both XPM and XGM nonlinear effects. We will discuss the idea and theory of this concept in section 4.1.3. After discussion of the operation principle, we show with experiment and simulation results how a return-to-zero (RZ) signal through this SOA-DI configuration is wavelength-converted into a non-return-to-zero (NRZ) signal at a speed as high as 160 Gbit/s. This speed is far exceeding the speed limitation due to the slow SOA recovery time.

4.1.1 Configuration and Description of a Delay Interferometer Filter

Configuration

The SOA-DI configuration is shown in Fig. 4.1. It comprises an SOA followed by a DI filter. A bandpass filter (BPF) is used between the SOA and the DI, but it can also be positioned after the DI. In such a configuration, an incoming signal P_{in} at a wavelength λ_{in} is used for gating the SOA-DI device. It is gated such that the device is open in the absence of the input signal but switched off when the input signal is launched into the device. A probe signal, usually a continuous wave (cw), on a new wavelength λ_{cnv} is used to visualize the gating operation. It will be shown that a fast switching is realized by using the DI filter.

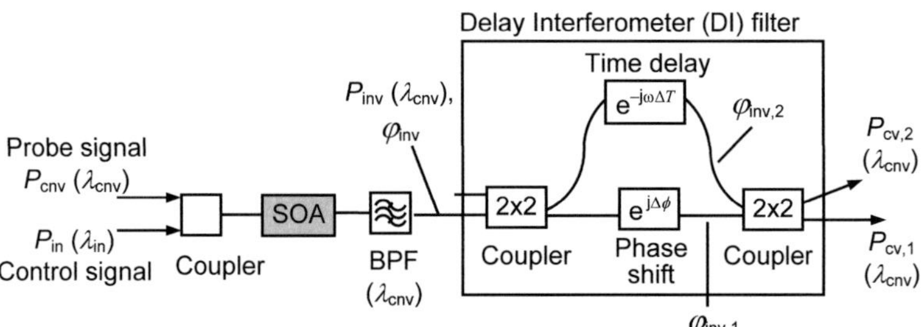

Fig. 4.1. All-optical wavelength converter based on an SOA followed by a delay interferometer (DI) filter. The input control signal P_{in} at a wavelength λ_{in} is converted to a signal P_{inv} at the wavelength λ_{cnv} after the SOA. P_{cnv} is reshaped by the DI-filter to a signal $P_{cv,1}$ and a signal $P_{cv,2}$. Their shapes depend on the parameters of the DI-filter, namely, the time delay ΔT and the phase shift $\Delta\phi$.

Transfer Function of a Delay Interferometer Filter

To understand the operation principle of the SOA-DI scheme, we start from the description of the DI filter in terms of transfer functions. For a DI filter shown in Fig. 4.1, we assume that two input signals to the lower and upper input ports of the DI filter are centered at the frequency f_0 and can be expressed by $A_{in,1}(t)\exp(j2\pi f_0 t)$ and $A_{in,2}(t)\exp(j2\pi f_0 t)$, respectively. $A_{in,1}(t)$ and $A_{in,2}(t)$ are the respective envelope functions. $A_{out,1}(t)$ and $A_{out,2}(t)$ are the envelope functions of the output signals at the lower and upper output ports. Thus, the signal P_{inv} in Fig. 4.1 corresponds to the signal $A_{in,1}$ and the signal $P_{cv,1}$ corresponds to the signal $A_{out,1}$. For these envelope functions, we define a baseband angular frequency Ω, which is the signal angular frequency $\omega = 2\pi f$ demodulated from the optical carrier angular frequency $\omega_0 = 2\pi f_0$

$$\Omega = 2\pi(f - f_0). \tag{4.1}$$

The input and output couplers are 2×2 directional couplers, whose bar power transmission is s (cross transmission $1-s$). The longer arm, i.e. the upper arm, of the delayed interference section provides a time delay of ΔT and the shorter arm, i.e. the lower arm, contains a tunable phase-shifter described by $e^{j\Delta\phi}$.

We can use an impulse response matrix $\underline{h}(t)$ to describe a DI filter. Following [42], we relate the input and output signals through a DI filter as

$$
\begin{pmatrix} A_{\text{out},2}(t) \\ A_{\text{out},1}(t) \end{pmatrix} = \underline{h}(t) * \begin{pmatrix} A_{\text{in},2}(t) \\ A_{\text{in},1}(t) \end{pmatrix}
$$

$$
= \int_{-\infty}^{\infty} \begin{pmatrix} \sqrt{s} & j\sqrt{1-s} \\ j\sqrt{1-s} & \sqrt{s} \end{pmatrix} \begin{pmatrix} \delta(\tau - \Delta T)e^{j2\pi f_0 \Delta T} & 0 \\ 0 & \delta(\tau)e^{j\Delta\phi} \end{pmatrix}
$$

$$
\times \begin{pmatrix} \sqrt{s} & j\sqrt{1-s} \\ j\sqrt{1-s} & \sqrt{s} \end{pmatrix} \begin{pmatrix} A_{\text{in},2}(t-\tau) \\ A_{\text{in},1}(t-\tau) \end{pmatrix} d\tau .
\tag{4.2}
$$

The first and third matrices on the right hand side of Eq. (4.2) are describing the input and output directional couplers, where a phase factor j is added to a signal that couples into the cross output. The diagonal elements of the middle matrix are the impulse response functions of the time delay function on the upper and of the phase shift $e^{j\Delta\phi}$ on the lower arm. Note that these two impulse response functions are defined with respect to the envelope functions[10]. A phase factor $j2\pi f_0 \Delta T$ is also added on the "long" arm, to take account of the additional phase shift on the optical carrier. The time delay ΔT of the "long" arm is adjusted such that ΔT approximates the bit period T_b, while $f_0 \Delta T$ is an integer number.

From Eq. (4.2) and with the power transmission of the coupler $s = \frac{1}{2}$ (i.e. a 3 dB coupler), the Fourier transforms of the output signal envelopes, $\breve{A}_{\text{out},1}(\Omega)$ and $\breve{A}_{\text{out},2}(\Omega)$, are then

$$
\begin{pmatrix} \breve{A}_{\text{out},2}(\Omega) \\ \breve{A}_{\text{out},1}(\Omega) \end{pmatrix} = \underline{H}(\Omega) \begin{pmatrix} \breve{A}_{\text{in},2}(\Omega) \\ \breve{A}_{\text{in},1}(\Omega) \end{pmatrix}
$$

$$
= \frac{1}{2} \begin{pmatrix} e^{-j\Omega\Delta T} - e^{j\Delta\phi} & je^{j\Delta\phi} + je^{-j\Omega\Delta T} \\ je^{j\Delta\phi} + je^{-j\Omega\Delta T} & e^{j\Delta\phi} - e^{-j\Omega\Delta T} \end{pmatrix} \begin{pmatrix} \breve{A}_{\text{in},2}(\Omega) \\ \breve{A}_{\text{in},1}(\Omega) \end{pmatrix},
\tag{4.3}
$$

where $\underline{H}(\Omega)$ is the transfer function matrix describing the frequency response of the DI filter, and $\breve{A}_{\text{in},1}(\Omega)$ and $\breve{A}_{\text{in},2}(\Omega)$ are the Fourier transforms of input signal envelopes,

$$
\breve{A}_{\text{out},1}(\Omega) = \frac{1}{2}\left[(e^{j\Delta\phi} - e^{-j\Omega\Delta T})\breve{A}_{\text{in},1}(\Omega) + (je^{j\Delta\phi} + je^{-j\Omega\Delta T})\breve{A}_{\text{in},2}(\Omega) \right],
\tag{4.4}
$$

$$
\breve{A}_{\text{out},2}(\Omega) = \frac{1}{2}\left[(je^{j\Delta\phi} + je^{-j\Omega\Delta T})\breve{A}_{\text{in},1}(\Omega) + (e^{-j\Omega\Delta T} - e^{j\Delta\phi})\breve{A}_{\text{in},2}(\Omega) \right].
\tag{4.5}
$$

---

[10]In the electrical engineering definition, the Fourier transform of an ideal time delay function $\delta(t-\Delta T)$ is $\exp(-j\omega\Delta T)$. The Fourier transform of $\delta(t-\Delta T)\exp(j\omega_0\Delta T)$ is then $\exp[-j(\omega-\omega_0)\Delta T] = \exp[-j\Omega\Delta T]$. With $\Omega = \omega - \omega_0$ and $\delta(\tau-\Delta T)A_{\text{in},1}(t-\tau) = A_{\text{in},1}(t-\Delta T)$, the Fourier transform of $A_{\text{in},1}(t-\Delta T)\exp[j\omega_0(t-\Delta T)]$ is

$$
\int A_{\text{in},1}(t-\Delta T)\exp[j\omega_0(t-\Delta T)]\exp(-j\omega t)dt = \int A_{\text{in},1}(t-\Delta T)\exp(-j(\omega-\omega_0)t)\exp(-j\omega_0\Delta T)dt
$$

$$
= \breve{A}_{\text{in},1}(\Omega)\exp(-j\Omega\Delta T)\exp(-j\omega_0\Delta T),
$$

and Fourier transform of $\exp(j\omega_0\Delta T)A_{\text{in},1}(t-\Delta T)\exp[j\omega_0(t-\Delta T)]$ is

$$
\exp(j\omega_0\Delta T)\int A_{\text{in},1}(t-\Delta T)\exp[j\omega_0(t-\Delta T)]\exp(-j\omega t)dt = \breve{A}_{\text{in},1}(\Omega)\exp(-j\Omega\Delta T).
$$

If the input signal $\breve{A}_{\mathrm{in},2}(\Omega)$ is zero, which is the case in Fig. 4.1, the transfer functions from the input signal $\breve{A}_{\mathrm{in},1}(\Omega)$ to the output signals are

$$H_{\mathrm{DI},1}(\Omega) = \frac{\breve{A}_{\mathrm{out},1}(\Omega)}{\breve{A}_{\mathrm{in},1}(\Omega)} = \frac{1}{2}(e^{\,j\Delta\phi} - e^{-j\Omega\Delta T}) = je^{\,j(\Delta\phi-\Omega\Delta T)/2}\sin\left(\frac{\Delta\phi+\Omega\Delta T}{2}\right), \tag{4.6}$$

$$H_{\mathrm{DI},2}(\Omega) = \frac{\breve{A}_{\mathrm{out},2}(\Omega)}{\breve{A}_{\mathrm{in},1}(\Omega)} = \frac{j}{2}(e^{\,j\Delta\phi} + e^{-j\Omega\Delta T}) = je^{\,j(\Delta\phi-\Omega\Delta T)/2}\cos\left(\frac{\Delta\phi+\Omega\Delta T}{2}\right). \tag{4.7}$$

The modulus square of respective transfer functions are connected to the power transmission.

The power transmission characteristics of a DI filter for various phase shifts and various time delays are depicted in Fig. 4.2(a) and (b), respectively. As the phase shift $\Delta\phi$ becomes $\pi/2$, the half of the center spectral component of the signal can pass the DI filter, seen in Fig. 4.2(a). As shown in Fig. 4.2(b), varying the time delay on the DI longer arm can give different transmission spectra. In the following, with the experiment and simulation results, we will explain how the DI filter mitigates the pattern effect in the SOA-based wavelength conversion.

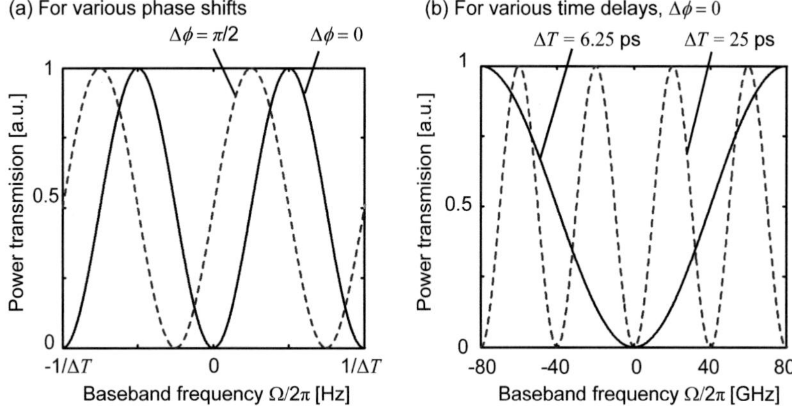

Fig. 4.2. Power transmission characteristics of a delay interferometer filter for (a) various phase shifts and (b) various time delays with a zero phase offset, where the *x*-axis is the envelope frequencies.

4.1.2 Operation Principle of Differential Delay Interferometer Scheme

The working principle is explained first in time domain. We assume that the optical signals experience mainly the cross-phase modulation (XPM). The input signal P_{in}, e.g. a pulse in Fig. 4.3(a), modulates the gain of the SOA and thereby the probe signal phase. The phase shift φ_{inv} of the output signal P_{inv} after the SOA is depicted in Fig. 4.3(b). The falling regime of the phase shift φ_{inv} is only limited by the pulse width of the P_{in} signal, whereas the recovery process of the φ_{inv} relaxes with the slower carrier recovery time. To suppress the slower recovery part of the signal after the SOA, the delay interferometer (DI) is used to perform a differential operation on this signal as follows. The signal P_{inv} is guided into the DI and split onto the two arms of the interferometer. The one on the longer arm of the DI is delayed by ΔT with respect to that on the shorter arm, before they are recombined in the output coupler of the DI. The respective phases $\varphi_{inv,1}$ and $\varphi_{inv,2}$ are shown in Fig. 4.3(c), where the phase difference $\varphi_{inv,1} - \varphi_{inv,2}$ is given Fig. 4.3(d). Note that the phase shift $\Delta\phi$ in Fig. 4.1 is assumed to be 0. In the absence of the input signal, the coupler directs the signal P_{inv} into the output port $P_{cv,2}$ in Fig. 4.1. In the presence of an input pulse, the signal on the lower arm carrying the phase shift $\varphi_{inv,1}$ first reaches the coupler. This phase shift $\varphi_{inv,1}$ (i.e. $\varphi_{inv,1} - \varphi_{inv,2} \neq 0$) opens the switching window and the coupler directs the signal P_{inv} into the output port $P_{cv,1}$. Later, when the signal on the upper arm reaches the coupler, the phase difference $\varphi_{inv,1} - \varphi_{inv,2}$ is reset to 0 and the switching window in the $P_{cv,1}$ port closes.

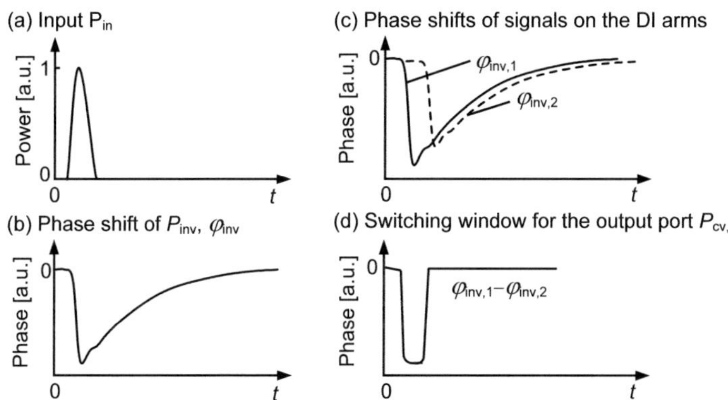

Fig. 4.3. **Forming a switching window in output port $P_{cv,1}$ of the delay interferometer, when a control pulse is launched into SOA. The input pulse P_{in} (a) induces a phase change φ_{inv} (b) on the output signal P_{inv} after the SOA. As shown in (c), the phase shift $\varphi_{inv,2}$ on longer arm is delayed with respect to the phase shift $\varphi_{inv,1}$ on shorter arm of the DI, Fig. 4.1. The difference between $\varphi_{inv,1}$ and $\varphi_{inv,2}$ forms a switching window for the signal through the DI.**

We now explain the working principle in frequency domain. From the spectral point of view, the transfer function of the SOA, also called as carrier density response (CDR) [63] to XGM and XPM, $H_{CDR}(\Omega)$ shows a kind of low-pass characteristic, seen in Fig. 4.4. This means that the frequency component at the optical carrier will be more amplified than the baseband signal, i.e. the modulation of the signal envelope. Note that only the blue part of the response functions (for the spectrum higher than the optical carrier) is shown in Fig. 4.4. Such a low-pass characteristic of the carrier density response can be understood as follows. As a control signal enters an SOA, it depletes the carriers in the SOA, and in turn the amplification of the probe signal is decreased. As the control signal dies out, a carrier recovery and a gain recovery for the probe signal take place. While the carrier depletion is a fast process, the gain recovery process is slow. The speed of the gain recovery is limited by the SOA carrier relaxation time. Now, for $\Delta\phi = 0$, the DI filter works as a differentiator for the signal in the output port $P_{cv,1}$, see Fig. 4.1. So the DI filter shows a high-pass characteristic as shown in Fig. 4.2(a). Note that only the right part of the DI filter in Fig. 4.2 is shown. As we add (in dB scale) the DI filter spectrum $|H_{DI}(\Omega)|$ on the SOA response function $|H_{CDR}(\Omega)|$, shown in Fig. 4.4, the transfer function of the wavelength converter is equalized. This explains the reason why the DI filter can enhance the performance of the SOA-based wavelength converters.

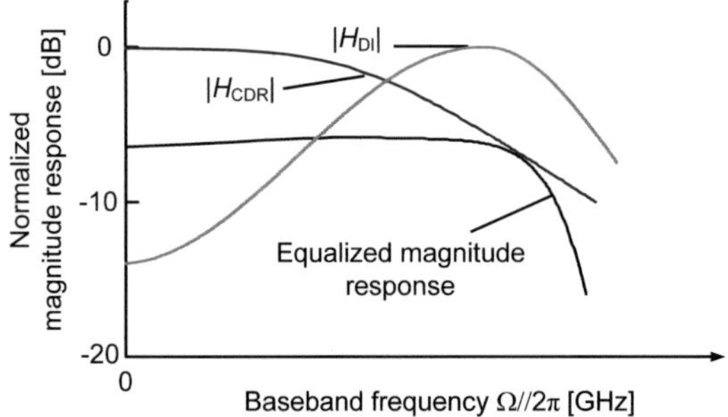

Fig. 4.4. Schematic description of the transfer function of an SOA $|H_{CDR}|$, also called as carrier density response, the transfer function of the DI filter $|H_{DI}|$, and the equalized magnitude response of the wavelength converter. Note that the *x*-axis is the baseband frequency.

In the above discussion for a zero phase shift $\Delta\phi$, the output signal $P_{cv,1}$ is non-inverted to the control signal. However, the switching windows for the output pulses in $P_{cv,1}$ are opened when there are input control pulses. The input pulses always induce gain saturation and lead to a low extinction ratio in the output signal. The output signal with an inverted[11] waveform has a good extinction ratio. To achieve an inverted signal, the phase offset of the DI is used for this purpose and will be discussed in next section.

4.1.3 Delay Interferometer Filter Used in Cross-Gain Modulation Based RZ-to-NRZ Wavelength Conversion

Published by J. Wang, A. Maitra, C. G. Poulton, W. Freude, and J. Leuthold, "Temporal dynamics of the alpha factor in semiconductor optical amplifiers," *IEEE J. Lightw. Technol.*, vol. 25, pp. 891-900, Mar. 2007.

As we discussed in section 4.1.2, the delay interferometer (DI) performs a differential operation on the cross-phase modulated (XPM) signal after the SOA, while cross-gain modulation plays a minor role. However, as the SOA operates at ultrafast speed e.g. 160 Gbit/s, the cross-gain modulation (XGM) on the one hand determines the performance of the wavelength

---

[11] The inverted operation means that a bit 1 (mark) is converted to a bit 0 (space), or vice versa.

converters. On the other hand, as a filter, the DI utilizes the XPM in the output signal after the SOA and performs a fast switching on the signal.

Very recently, an SOA based all-optical wavelength conversion experiments with real signal regeneration was demonstrated. In the corresponding reference [47], the amplitude and phase responses for a short bit sequence were also recorded. With the model from Chapter 2, we now explain the operation principle and the experimental results as follows. It is noticed that the experiment and the theory are matched to a degree not possible before, even under situations where pattering occurs.

The wavelength converter comprises an SOA followed by a delay interferometer (DI), Fig. 4.1(a). The SOA had an active layer length of 2.6 mm, and had the gain and phase recovery dynamics similar to the device shown in Fig. 3.2 and Fig. 3.14. The delay in the interferometer is $\Delta T = 6.25$ ps and matches the duration of one bit.

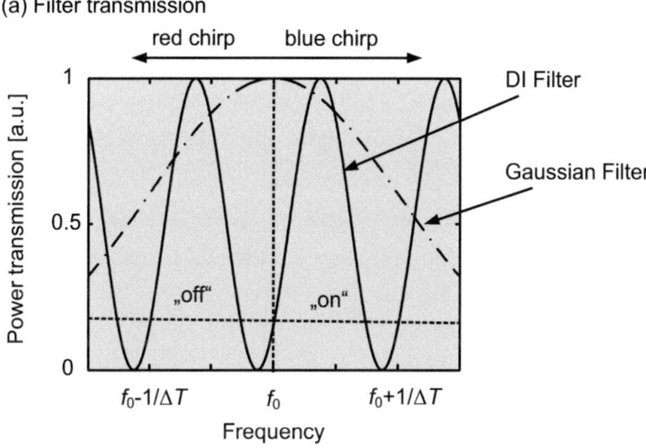

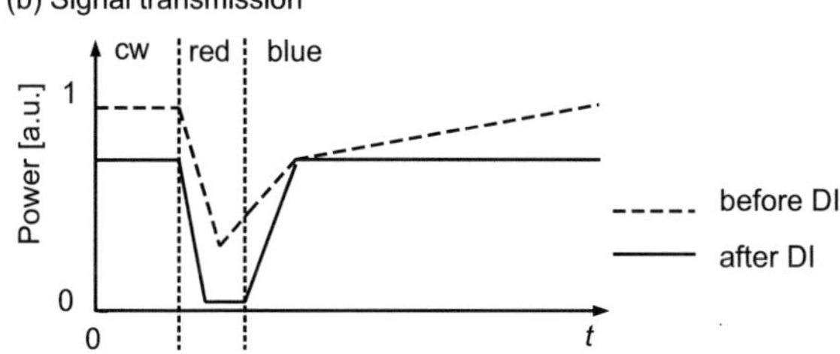

Fig. 4.5. (a) Transmission window of the delay interferometer (DI) filter and the Gaussian filter (note that the position of the Gaussian filter in [82] was wrong). For a good "on" state, the phase ($\Delta\phi$) of the DI filter is detuned such, that the minimum of the transmission window is offset by 45 degrees from the cw-frequency f_0, where T is the duration of one bit. The converter is switched "off", when the gain of the SOA is suppressed due to XGM or the inverted signal is strongly red-chirped on the falling slope. As the cw frequency component and the blue-chirped part are equalized by this DI, the converted signal after the SOA (before the DI), dashed line in (b), is reshaped to the signal depicted as solid line in (b).

Fig. 4.5(a) shows the power transmission characteristic of the DI versus the optical frequency. The device is operated in an inverting mode. A bit "1", high power level, or a series of bit "1" introduced into the device is actually converted into a bit "0", low power level, or a series of bit "0"s by means of the XGM and XPM effect. The converter is switched off. If a "0" is launched into the device, the device switches back into the "ON" state. This turning-on needs to happen fast and should not be limited by the slow SOA carrier recovery time. For improving the speed of the device we use a trick. The trick is as follows. In its switched "ON" state, the device is not fully switched on. Actually, the DI filter is detuned in such a way that in its switched "ON" state only 20% of the total power of the cw signal at f_0 transmits. Through the DI filter, the transmissions of the cw frequency component and the blue-chirped part are equalized, while the red-chirped part is blocked. The converted signal after the SOA (before the DI), dashed line in Fig. 4.5(b), is reshaped to the signal depicted as solid line in Fig. 4.5(b). The details of the scheme will be explained below.

For the experiment, we have launched a "0001011010000000" sequence into the device. The measured output power of the converted signal is depicted as a solid line in Fig. 4.6(d). The experiment shows a nice NRZ signal pulse shape with steep transients between "0"s and "1"s. These steep transients are hard to interpret because the device has a much longer carrier recovery time. As shown in Fig. 4.6(a), the output signal after the SOA has neither good low-level "0"s nor well-formed high-level "1"s.

Yet, the amplitude and phase response shown in Fig. 4.6(a) and (b) show the origin of this favorable behavior, while Fig. 4.6(c) also gives the temporal evolution of the α_{eff}-factor and the chirp experienced by the CW signal in the SOA. At first the device is switched "on" and 20% of the cw signal passes through the DI. When the device is switching from "on" to "off", then the amplitude is suppressed due to fast XGM effects, see Fig. 4.6(a). The device switches off. Yet, the switching off does not last long, since XGM is dominated to a considerable extent by the fast gain compression effects which only last for 1 to 2 ps. Now with a 2 ps delay a negative phase shift sets in, see Fig. 4.6(b), accompanied with a red-chirp, Fig. 4.6(c). From the position of the filter in Fig. 4.5(a), one now can see that a red-chirp on top of a cw frequency at f_0 detunes the center frequency to the left – where the transmission of the DI filter is in its "OFF" state. After the pulse has passed, the device switches back to the "ON" state. This in part happens due to the gain recovery. To another part this relaxing back to the "ON" state is strongly supported by the phase recovery. Fig. 4.6(e) shows that the phase switches back and even overshoots. From Fig. 4.6(c) one can see that this overshooting really leads the device into the blue-chirped region where the DI filter is open.

Fig. 4.6(d) and (e) also show that the simulations (dashed lines) are in good agreement with the experiments (solid lines). For the simulations we used the same parameters as in the experiment reported in [47], also given in Appendix B. For instance, the SOA had an active layer length of 2.6 mm. The average power of the pulse train over 100 ps was 2.9 dBm, while the cw signal had a power level of 12.6 dBm. The splitting ratio of the optical coupler before the SOA was 30:70 for the cw and the pulse train signal, respectively. Other parameters are given in Appendix B.

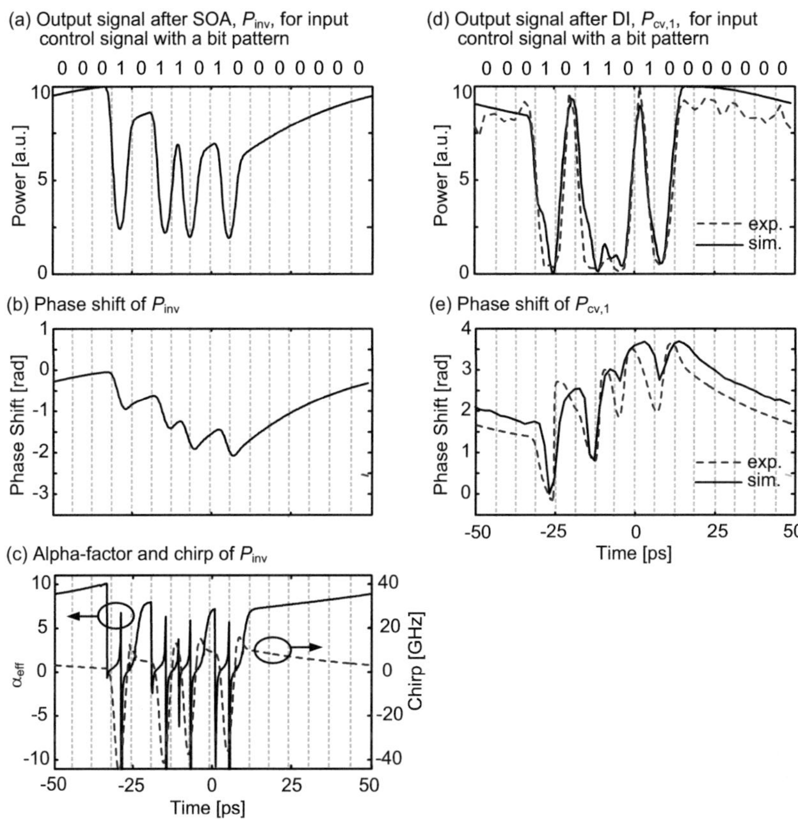

Fig. 4.6. Wavelength converted bit-sequence. (a) and (d) are power transmissions after the SOA and after the DI filter. (b) and (e) are phase changes after the SOA and after the DI filter. The input bit sequence was 0001011010000000. In (d) and (e) we show measured (solid) and calculated (dashed) curves. The plots in (c) show the calculated α_{eff} and chirp of the converted signal after the SOA but before DI.

As mentioned before, only 20% of the total power of the cw signal at f_0 transmits through the DI. The remaining 80% of the power are used to overcome pattern effects. To illustrate this point, we have plotted the simulated output powers at two output ports in Fig. 4.7. We see that clear level "1"s appear in the output port $P_{cv,1}$, while most power is guided into the output port $P_{cv,2}$. Also, as $P_{cv,1}$ signal has as low as possible power at level "0"s, the residual power is guided into the port $P_{cv,2}$.

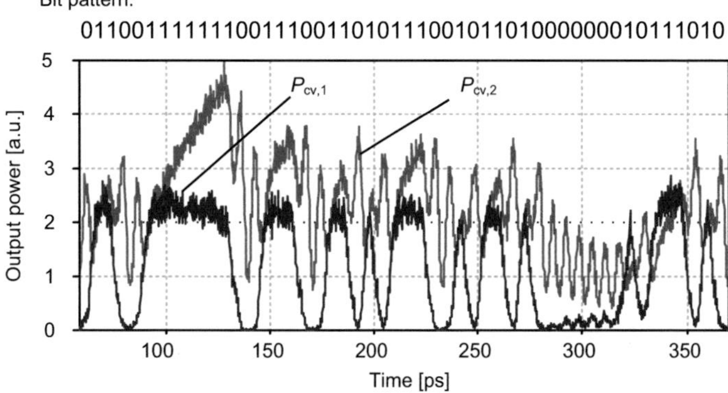

Fig. 4.7. Simulated 160 Gbit/s output signals at two output ports of a DI filter. The output signal $P_{\mathrm{out,1}}$ is obviously a NRZ signal. The DI filter has a time delay of one bit delay, namely 6.25 ps, and a phase shift of $\pi/4$.

We note that an inverted NRZ signal will not be available, when the time delay ΔT is smaller with respect to the pulse width. With a time delay ΔT of one half or one bit duration, we performed the system simulations for the wavelength conversion at 160 Gbit/s. The input RZ pseudo random bit sequence (PRBS) has a length of $2^{12}-1$. To qualify the output signal, we use a signal quality factor, i.e. Q-factor. Considering the optimum decision threshold, the Q-factor (in dB) is defined as

$$Q_{\mathrm{lin}} = \frac{m_1 - m_0}{\sigma_1 + \sigma_0}, \quad Q = 10\log_{10}\left(Q_{\mathrm{lin}}\right)^2 = 10\log_{10}\left(\frac{m_1 - m_0}{\sigma_1 + \sigma_0}\right)^2, \tag{4.8}$$

where m_1 and m_0 are the mean optical powers at high (bit "1") and low (bit "0") levels, σ_1 and σ_0 are the standard deviation of the respective optical powers. Sometimes, the Q-factor (in dB) is also called as Q^2-factor, since it is obtained by taking the square of the linear Q_{lin}-factor. The eye diagram and Q-factor of the inverted signal after the SOA are shown in Fig. 4.8(a). The signal after the SOA has a bad quality. The eye diagrams of converted signals after the DI having respective time delays are compared in Fig. 4.8(b) and (c). Both, with the DI having a time delay ΔT of 3.125 ps or 6.25 ps, converted signals at 160 Gbit/s have open eyes and thus good quality. In other words, the pattern effect originating from the SOA carrier dynamics can be mitigated by applying the DI. However, the signal in Fig. 4.8(b) is inverted but "return-to-one", while the signal in Fig. 4.8(c) is inverted and NRZ.

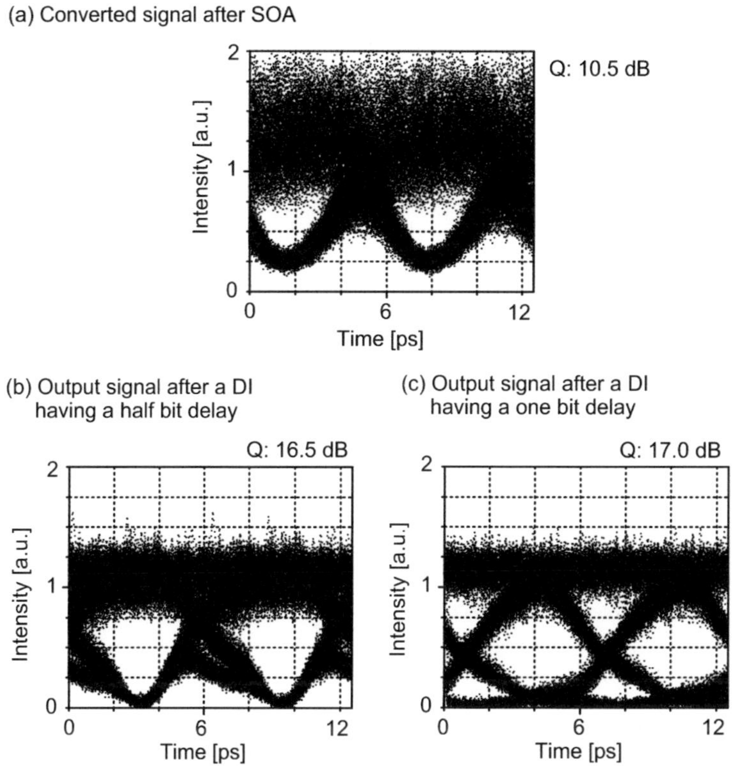

Fig. 4.8. Eye-diagrams of simulated wavelength converted signals at 160 Gbit/s after a SOA-DI wavelength conversion scheme. (a) shows the signal after the SOA, (b) and (c) show the output signals after the DI filter having a half bit delay and a one bit delay, namely 3.125 ps and 6.25 ps, respectively. While the signal in (b) is an inverted "RZ" with opened eye, the signal in (c) is an inverted NRZ signal with opened eye. The phase shifts of the DI in both cases are 45 degree.

As a conclusion for this section, an SOA-DI based wavelength conversion has been investigated. We showed that, by properly detuning the differential time delay in the DI, a RZ to NRZ wavelength conversion can be achieved at high speed. This is achieved by utilizing the relative temporal lag of the XPM, which is comparable with respect to the pulse width.

4.2 Pattern Effect Mitigation Using a Pulse Reformatting Optical Filter

Since the delay interferometer is, in fact, just a filter, it is interesting to extend the investigation to see if other types of filters can perform similar operations. This type of study has been presented for a fiber Bragg grating (FBG) [91], which optimized the frequency response of an SOA-based wavelength converter. Recently, wavelength conversion schemes using an SOA followed by a single red-shift optical filter [44], or a blue-shift optical filter [49] and [62] or finally a pulse reformatting optical filter (PROF) [46] have been introduced.

The PROF scheme demonstrated wavelength conversion at 40 Gbit/s in [46], with record low input data signal powers of −8.5 and −17.5 dBm for non-inverted and inverted operation. This is almost two orders of magnitudes less than typically reported for 40 Gbit/s wavelength conversions. The reason for the good conversion efficiency lies in the design of the filter. The scheme explicitly takes into account both SOA XGM and XPM effects. Indeed, this scheme provides the best possible conversion efficiency for an SOA-based wavelength converter or regenerator. The PROF scheme can offer advantages as follows:

- Simultaneous all-optical wavelength conversion, switching, power equalization and output format conversion.

- Possible multi-wavelength conversion of WDM signals, if a multiplexer and a channel interleaver are used as [44].

- Transparency to bit-rates as long as the filter has the passband for both low and high speed and the speed limitation of the nonlinear element is not exceeded.

4.2.1 Configuration and Operation Principle

Configuration

The all-optical wavelength converter comprises an SOA and a pulse reformatting optical filter (PROF) schematically shown in Fig. 4.9. For all-optical wavelength conversion, the input data signal P_{in} and the probe signal P_{cnv} are launched into the SOA. In the SOA, the input data signal encodes the signal information by means of XGM and XPM onto the probe signal. As a result, we obtain at the SOA output a chirped and inverted signal P_{inv} at the wavelength of the probe signal λ_{cnv}. The purpose of the subsequent passive optical filter with appropriate amplitude and phase responses is to reformat the signal P_{inv} into a new output-signal P_{cv} with a determined shape. However, the inverted signal normally has a strong pattern effect. Hence, in order to reduce the pattern effect and convert the inverted signal into the non-inverted one, an optimum optical filter should be designed and then approached.

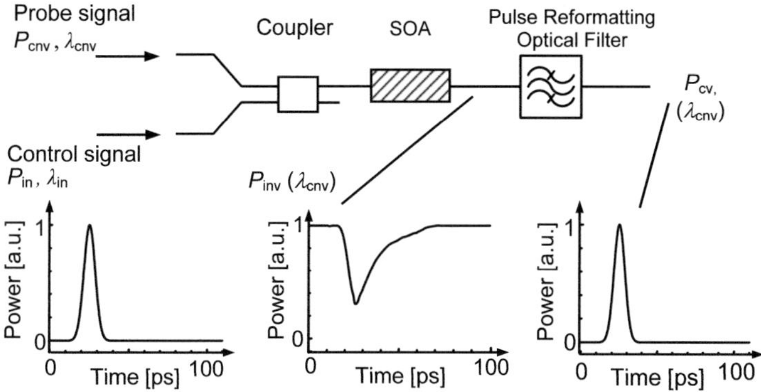

Fig. 4.9. All-optical wavelength converter comprising an SOA and a pulse reformatting optical filter (PROF). In this scheme, the input signal P_{in} is mapped by means of SOA cross-gain and cross-phase modulation into an inverted pulse P_{inv} at the probe wavelength. A filter, namely PROF, then reformats the signal P_{inv} into an output signal with the desired pulse shape of P_{cv}.

Description of a Pulse Reformatting Optical Filter

A linear filter, such as a PROF, can be described by the frequency response function $H_{\text{PROF}}(\Omega)$, where Ω is the baseband angular frequency,

$$H_{\text{PROF}}(\Omega) = |\, H_{\text{PROF}}(\Omega)\,| \exp\left[j \varphi_{\text{PROF}}(\Omega) \right], \tag{4.9}$$

$$\varphi_{\text{PROF}}(\Omega) = \arctan\left(\frac{\text{Im}(H_{\text{PROF}}(\Omega))}{\text{Re}(H_{\text{PROF}}(\Omega))} \right). \tag{4.10}$$

$|\, H_{\text{PROF}}(\Omega)\,|$ and $\varphi_{\text{PROF}}(\Omega)$ stand for the spectral magnitude and phase of filter. According to Fig. 4.9, the transfer function of the PROF can be calculated by applying optical filter theory [51]. Accordingly, the Fourier transform of the target output signal $\breve{A}_{\text{cv}}(\Omega)$ (corresponding to P_{cv} in Fig. 4.9) is divided by the Fourier transform of the inverted signal $\breve{A}_{\text{inv}}(\Omega)$ (corresponding to P_{inv} in Fig. 4.9) after the SOA,

$$H_{\text{PROF}}(\Omega) = \frac{\breve{A}_{\text{cv}}(\Omega)}{\breve{A}_{\text{inv}}(\Omega)}. \tag{4.11}$$

Another important parameter of a linear filter is group delay τ_g, which is defined as the first order derivation of the spectral phase with respect to frequency

$$\tau_g = -\frac{d\varphi_{\text{PROF}}(\Omega)}{d\Omega}. \tag{4.12}$$

From the physical point of view, and particularly in optics, the group delay has the meaning that it is the rate of change of the total phase with respect to frequency.

We have simulated a wavelength conversion in the SOA with a 40 Gbit/s pseudo random bit sequence (PRBS) control signal P_{in} and got an inverted signal P_{inv}, whose spectrum is shown in Fig. 4.10(a). By applying Eq. (4.11) to a desired 40 Gbit/s PRBS signal P_{cv}, whose spectrum is shown in Fig. 4.10(b), a typical PROF power transmission $|H_{PROF}(\Omega)|^2$ is shown as solid line in Fig. 4.10(c). The corresponding group delay τ_g is depicted as solid line in Fig. 4.10(d).

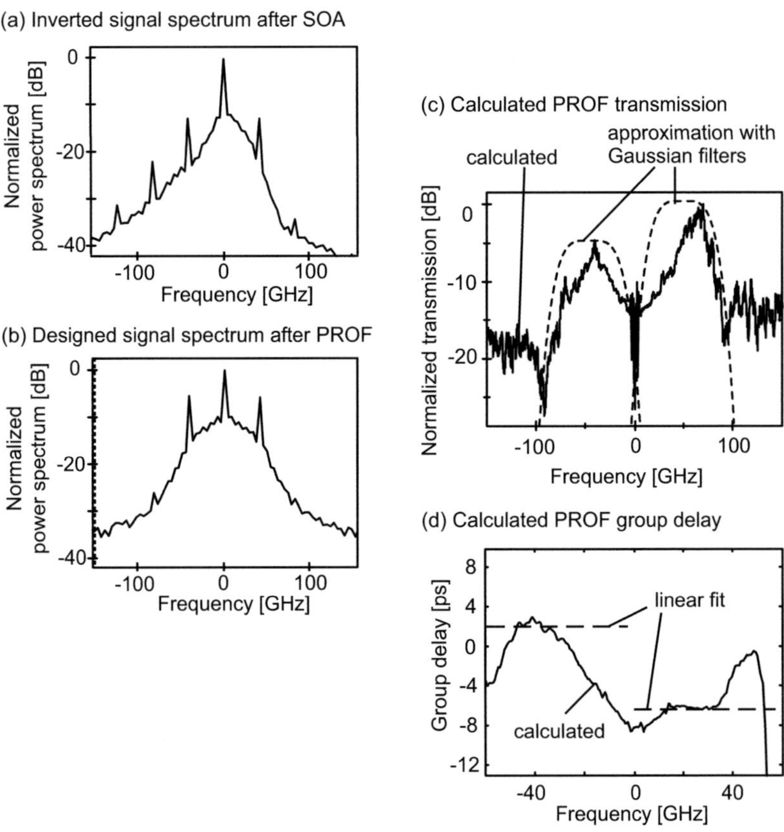

Fig. 4.10. Power spectra of (a) an inverted PRBS signal after an SOA and (b) a desired PRBS signal. By applying Eq. (4.11), an ideal PROF transmission is gotten and plotted as a solid line in (c), where the dashed line is an approximation with two Gaussian filters. (d) is the calculated group delay of the PROF (solid line) and its linear fit (dashed line).

In Fig. 4.10(c) and (d), we can see that the calculated filter transmission and group delay have complicated shapes and are not easily realizable. Nevertheless, we can approximate intuitively such a calculated PROF by means of two individual filters as depicted by the dashed lines in Fig. 4.10(c). In this approximation we cut the frequency spectrum into two spectral slices. Individual attenuation and group delay (shown as dashed line in Fig. 4.10(d)) to each of the spectral slices are introduced before recombining two spectral slices into a single synthesized filter.

The above discussion allows us to analyze the principal of PROF from the physical point of view. The physical processes taking place in the SOA are not only the XGM but also the XPM. As seen in Fig. 3.5, the XPM can be split into a red-chirped and a blue-chirped part. According to the calculated PROF transmission and its group delay, what the PROF does actually is splitting the red-chirped part away from the blue-chirped part. Then, we can get the reconstructed red-chirped signal and blue-chirped signal separately. Finally, the blue-chirped signal is combined with the red-chirped signal which is delayed by the group delay time. Indeed, the best result is obtained when the red-chirped and blue-chirped signals have the similar amplitudes. The group delay can not be neglected because different delay times correspond to different phase shifts. As a matter of fact, we change the time delay in order to change the red and blue part spectrums. Therefore after a mode-beating[12] (spectrum superposition) configuration for red and blue part spectral components, we can regulate the shape of the combined signal and finally get the optimized reconstructed signal

Realization Scheme

Following the general designing concept of a PROF, we can realize it by using a red-shift optical filter (RSOF) and a blue-shift optical filter (BSOF) and including proper relative attenuation and time delay between two filters, shown in Fig. 4.11.

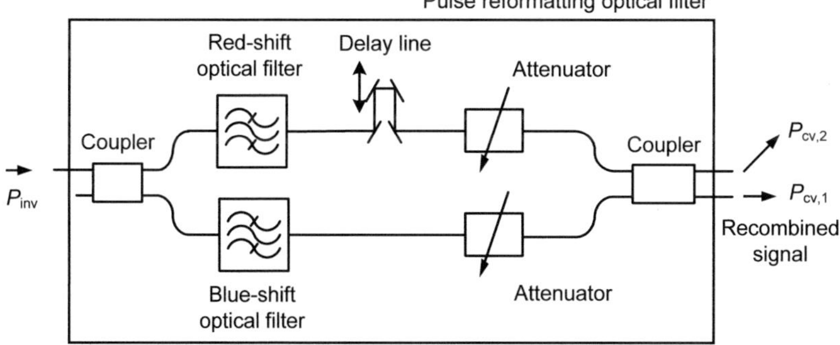

Fig. 4.11. Realization scheme of a pulse reformatting optical filter (PROF), reshaping an inverted signal P_{inv} after an SOA to a recombined signal with desired shape $P_{cv,1}$. The PROF consists of a red shifted optical filter and a blue shifted optical filter. A delay line and attenuators are used to construct a good recombined signal.

For such a scheme in Fig. 4.11, the transfer function of the PROF is expressed by using the transfer functions of the respective filters and adding the proper relative attenuation and

---

[12] Note that the terminology of mode beating used in [46] is not referring to the beating between different propagation modes. In fact, this terminology is referring to the superposion of different spectral components.

time delay between two filters. We define the transfer functions of the RSOF and the BSOF as $H_R(\Omega)$ and $H_B(\Omega)$

$$H_R(\Omega) = |H_R(\Omega)| \exp[j\varphi_R(\Omega)], \quad H_B(\Omega) = |H_B(\Omega)| \exp[j\varphi_B(\Omega)], \tag{4.13}$$

where $|H_R|$ and $|H_B|$ are the spectral magnitude, φ_R and φ_B are the spectral phase of the respective filters. The attenuators after the RSOF and the BSOF have power attenuations of $(A_R)^2$ and $(A_B)^2$, respectively, where $A_{R(B)} \leq 1$. The delay line on the RSOF arm induces a time delay of Δt corresponding a Fourier transform $\exp(-j\Omega\Delta t)$ with respect to baseband frequency. Assuming that the output signal after the SOA is $\breve{A}_{\mathrm{inv}}(\Omega)$ and two couplers are 3 dB couplers, the output signals of the PROF can be written as

$$\begin{pmatrix} \breve{A}_{\mathrm{cv},2}(\Omega) \\ \breve{A}_{\mathrm{cv},1}(\Omega) \end{pmatrix} = \begin{pmatrix} \dfrac{\sqrt{2}}{2} & \dfrac{\sqrt{2}}{2}j \\ \dfrac{\sqrt{2}}{2}j & \dfrac{\sqrt{2}}{2} \end{pmatrix} \begin{pmatrix} A_R H_R(\Omega)\exp(-j\Omega\Delta t) & 0 \\ 0 & A_B H_B(\Omega) \end{pmatrix}$$

$$\times \begin{pmatrix} \dfrac{\sqrt{2}}{2} & \dfrac{\sqrt{2}}{2}j \\ \dfrac{\sqrt{2}}{2}j & \dfrac{\sqrt{2}}{2} \end{pmatrix} \begin{pmatrix} \breve{A}_{\mathrm{inv}}(\Omega) \\ 0 \end{pmatrix}. \tag{4.14}$$

The transfer function of the PROF filter between the output signal $\breve{A}_{\mathrm{cv},1}(\Omega)$ (recombined signal $P_{\mathrm{cv},1}$ in Fig. 4.11) and the input signal $\breve{A}_{\mathrm{inv}}(\Omega)$ (inverted signal P_{inv} in Fig. 4.11) is then

$$H_{\mathrm{PROF}}(\Omega) = |H_{\mathrm{PROF}}(\Omega)| \exp[j\varphi_{\mathrm{PROF}}(\Omega)]$$

$$= \frac{\breve{A}_{\mathrm{cv},1}(\Omega)}{\breve{A}_{\mathrm{inv}}(\Omega)} = \frac{j}{2}[A_R H_R(\Omega)\exp(-j\Omega\Delta t) + A_B H_B(\Omega)]. \tag{4.15}$$

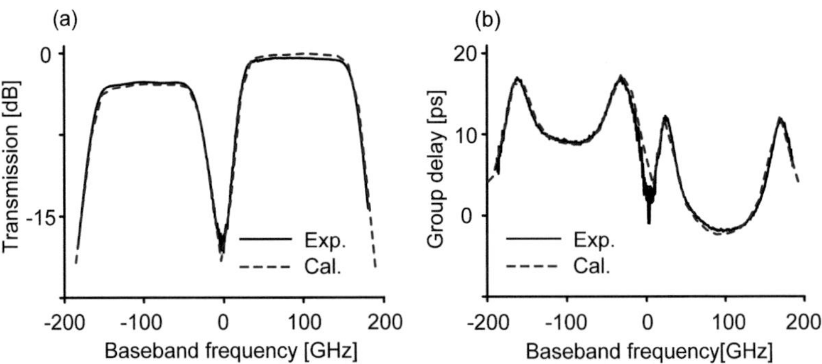

Fig. 4.12. Calculated PROF transmission and group delay, solid lines in (a) and (b), by applying Eq. (4.15) with the measured transfer functions of BSOF and RSOF and the fitted relative attenuation and group delay between BSOF and RSOF. Calculation results coincide with the experiment results, dashed lines in (a) and (b).

To calculate the transfer function of the PROF shown in Fig. 4.11, we can substitute the measured transfer functions of RSOF and BSOF into Eq. (4.15). We can also apply Eq. (4.15) to estimate the relative power attenuation $20 \cdot \log_{10}(A_B/A_R)$ and the relative group delay Δt between the BSOF and the RSOF. An example result is shown in Fig. 4.12, where the calculation results (solid lines) match the experimental results (dashed lines) quite well.

Equalizing of Frequency Response

Now we explain the operation principle of the PROF in the frequency domain by means of an equalization of frequency response.

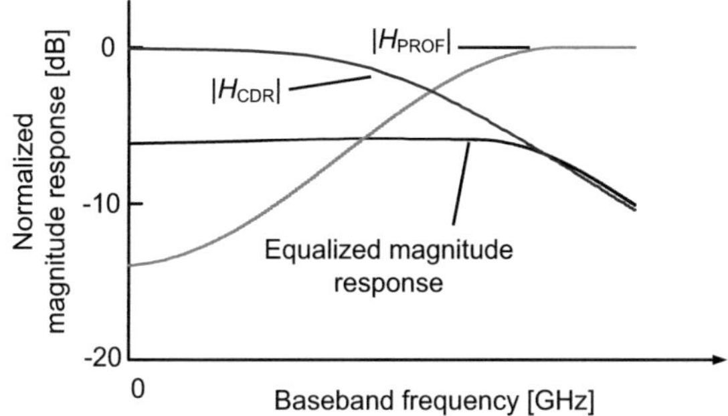

Fig. 4.13. Schematic description of the transfer function of an SOA $|H_{\mathrm{CDR}}|$, also called as carrier density response, the transfer function of the PROF filter $|H_{\mathrm{PROF}}|$, and the equalized magnitude function of the wavelength converter. The x-axis is the modulation frequency. Note that only the right parts of the transfer functions are given.

As discussed in former section, the carrier density response (CDR) $H_{\mathrm{CDR}}(\Omega)$ of the SOA to XGM and XPM shows a kind of low-pass characteristic, seen in Fig. 4.13. This means that the frequency component at the optical carrier will be more amplified than the sidebands, i.e. the modulation of the baseband signal. Note that only the blue part of the response functions is shown in Fig. 4.13. Therefore, the task of a PROF is actually to suppress the transmission of the optical carrier and simultaneously equalize the low-pass characteristic of CDR. The result of the equalization means a flat frequency response for the whole wavelength converter, including the SOA, seen in Fig. 4.13.

Mode Beating (Spectrum Superposition)[13]

The PROF filter concept can also be understood intuitively by means of mode beating. In Fig. 4.10, we have plotted a linear fit for the calculated group delay, since we concern ourselves with the group delay difference not the exact group delay values of red and blue-shifted part. In accordance with Fig. 4.11, we split off the red-chirped from the blue-chirped part, delay the leading red-chirped part by a time delay of Δt with respect to the trailing blue

---

[13] See footnote 12.

chirped part, and then recombine them. The effect of splitting off and recombining the two spectral components while introducing different time delays are plotted schematically in Fig. 4.14. Fig. 4.14(a) to (c) show the combined signal (solid line) obtained when recombining the red chirped (dashed line) and the blue-chirped (dotted line) signals after time delays of 0 ps, 10 ps and 20 ps. The best result is obtained when the red and blue-chirped signals have similar amplitudes and when the red chirped signal is delayed by 10 ps. Note that the 10 ps time delay is in accordance with the calculated group delay in Fig. 4.10(d). Fig. 4.14 shows that the scheme given in Fig. 4.11 is as a matter of fact a mode-beating (spectrum superposition) configuration for the red and blue spectral components.

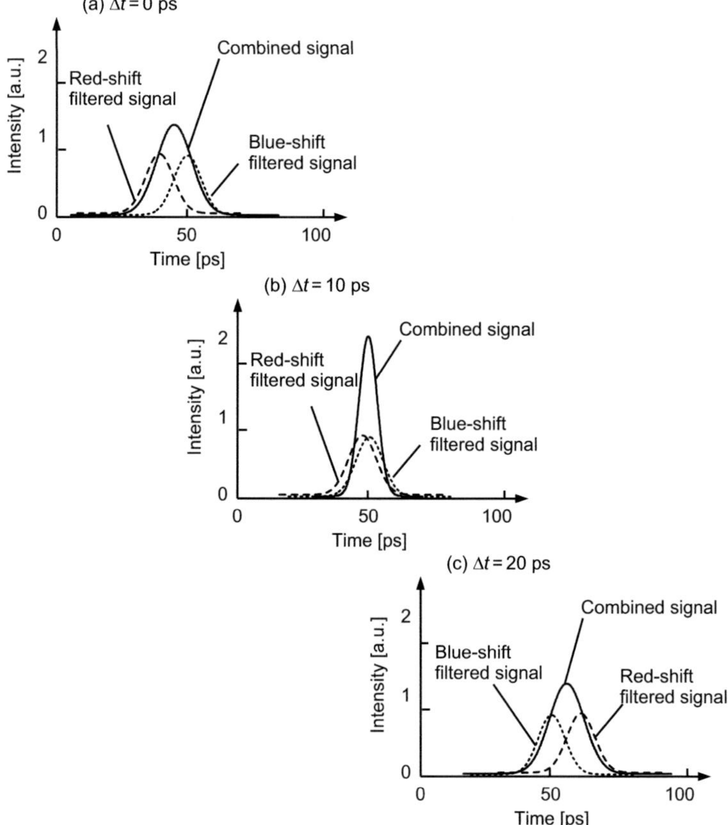

Fig. 4.14. Illustration of the mode beating (spectrum superposition). The split-off red chirped (dashed) and blue chirped (dotted) spectral parts of the inverted signal are recombined with time delays of 0, 10 and 20 ps. A new signal (solid line) is produced. An almost ideal RZ signal with narrow Full-Width Half Maximum is obtained if the leading red-chirped spectrum is delayed by 10 ps with respect to the blue spectral component.

4.2.2 Pattern Effect Mitigation Experiment

Published by J. Wang, A. Marculescu, J. Li, P. Vorreau, S. Tzadok, S. Ben Ezra, S. Tsadka, W. Freude, J. Leuthold, "Pattern effect removal technique for semiconductor optical amplifier-based wavelength conversion," *IEEE Photon. Technol. Lett.*, vol. 19, no. 24, pp. 1955-1957, Dec. 2007. / Proc. 33th European Conf. Opt. Commun. (ECOC'07), paper Tu3.4.6, Berlin, Germany, Sept. 16-20, 2007.

The PROF scheme basically represents an optimum filter for the SOA response with the potential for highest speed operation [63]. An experiment implementing a PROF based on MEMS technology demonstrated wavelength conversion at 40 Gbit/s [46], with record low input data signal powers of −8.5 and −17.5dBm for non-inverted and inverted operation. However, while the PROF scheme has been shown to work, it is so far not clear if the scheme can overcome pattern effects so that it might be used at highest speed as well.

In this section, we will show that the PROF scheme indeed and effectively mitigates SOA pattern effect. The pattern effect mitigation technique demonstrated here is based on the fact that the red chirp (decreasing instantaneous frequency) and the blue chirp (increasing instantaneous frequency) in the inverted signal behind an SOA have complementary pattern effects. If the two spectral components are superimposed by means of the PROF, then pattern effects can be successfully suppressed. An experimental implementation at 40 Gbit/s shows a signal quality factor improvement of 7.9 dB and 4.8 dB if compared to a blue or red shifted optical filter assisted wavelength converter scheme, respectively. This technique is also supported by simulation.

Pattern Effect Removal Experiment at 40 Gbit/s

The scheme for the all-optical wavelength conversion with simultaneous pattern effect removal is shown in Fig. 4.15. The setup comprises an SOA followed by a pulse reformatting optical filter (PROF). The current PROF is built with discrete components, so that any parameters can be detuned and optimized individually. The PROF consists of a blue shifted optical filter (BSOF) and a red shifted optical filter (RSOF). An OD and a VOA were added after the RSOF. A conventional band-pass filter (BPF) has been added at the output of the PROF to curtail the spectrum of the converted signal to the ITU channel passband.

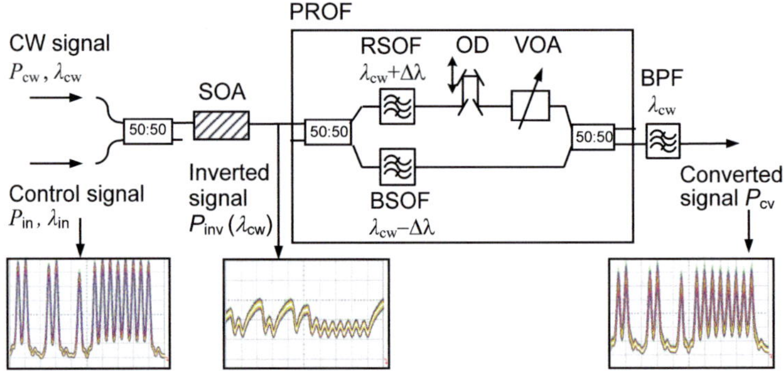

Fig. 4.15. Wavelength conversion scheme with an SOA followed by a pulse reformatting optical filter (PROF), which transfers an inverted signal with strong pattern effect to non-inverted and pattern dependency removed signal. OD: optical delay, VOA: variable optical attenuator.

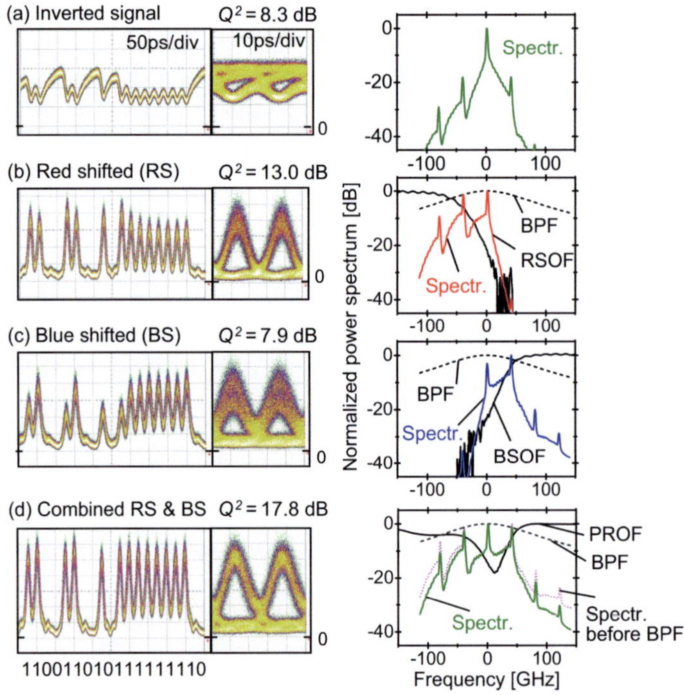

Fig. 4.16. Wavelength conversion experiment results, bit pattern (left column), eye diagram (middle column) and spectra with the filter shape (right column). Spectra are centered at the CW carrier frequency f_{cw}. (a) Inverted converted signal after SOA, (c)-(d) RS, BS and combined signal.

The wavelength conversion with simultaneous pattern effect removal works as follows. A data signal modulates both the gain and the refractive index of the SOA, thereby impressing the information in an inverted manner onto another continuous wave (cw) signal. Fig. 4.16(a) shows the pattern and power spectrum of the inverted cw signal behind the SOA. The leading edges of the converted light pulses are spectrally red-shifted (RS) whereas the trailing edges are blue-shifted (BS). In our scheme we now split off the RS and the BS spectral components. The respective pattern and spectra are depicted in Fig. 4.16(b) and (c). It can be noticed that the two signals have complimentary pattern effects. The signal qualities of the respective signals are 13.0 and 7.9 dB. While the RS pattern shows an overshoot in the first bit in a sequence of "1"s, the BS pattern undershoots particularly the first bit in a long sequence of "1"s. When the two signals are combined – after introducing a proper delay and attenuation onto the RS signal – the two complimentary patterns compensate to each other, resulting in a non-inverted signal with a signal quality of as much as 17.8 dB. The pattern, eye diagram and the spectrum of the combined, wavelength converted signal are shown in Fig. 4.16(d).

In the experiment, a 33% RZ data signal at a bit rate of 40 Gbit with an average power P_{in} between -5 and 12.7 dBm and a carrier wavelength $\lambda_{in} = 1530$ nm is launched into the SOA. The pseudo-random data with a sequence length of $2^{11} - 1$ has a quality factor $Q^2 = 19.8$ dB. A cw signal with P_{cw} between -4 and 16 dBm and $\lambda_{cw} = 1536.1$ nm is coupled into the SOA. It becomes the wavelength converted signal. The bulk SOA of length $L = 2.6$ mm was biased at $I = 750$ mA. The non-saturated and saturated gain was 23 and 3 dB. The SOA recovery time (10% to 90%) is ~45 ps and the polarization sensitivity is 0.5 dB. The plots in Fig. 4.16 have been recorded with a signal input power of 12.7 dBm and a cw power of 16 dBm. However, the input dynamic range of the device is large and similar results (Q^2-factors of 16 dB) were obtained with input and cw signals of -4 dBm. We will discuss this result later. This is not yet as good as in [46] but also, the current setup is built with lossy discrete components.

For this experiment we used two thin-film optical filters to realize the PROF. The parameters of the detuned filters are given in Table 4.1, and the filter passbands are also depicted in Fig. 4.16(b) and (c).

Table 4.1. Parameters of filters used. PMD: polarization mode dispersion.

Filter	Center λ [nm]	0.5 dB bandwidth	30 dB bandwidth [nm]	PMD [ps]
RSOF	1536.683	1.002	2.121	≤ 0.10
BSOF	1535.352	1.035	2.148	≤ 0.10

The optimum mode beating between red- and blue-chirped signals is obtained when two signals have similar amplitudes and the time delay between the red- and blue-chirped signals vanishes [46]. In our case, we attenuated and delayed the red-chirped signal by ~5 dB and ~10 ps with respect to the blue-chirped signal. The transfer function of the optimized PROF has been measured. In Fig. 4.17(a) and (b), the transmission spectrum and group delay spectrum are depicted as solid lines. The ideal PROF filter transmission spectrum as derived from a calculation (Eq. (4.11)) is plotted into the same figure (dashed lines). The PROF filter used in the experiment and the ideal filter are fairly similar. The Q^2-factors reported in Fig. 4.16 were recorded for the best combined signal. In our case, by varying the position of the filters, the Q^2-factor of the BS signal can be as good as 9.9 dB, while the Q^2-factor of the RS signal can not be improved beyond the 13dB reported above.

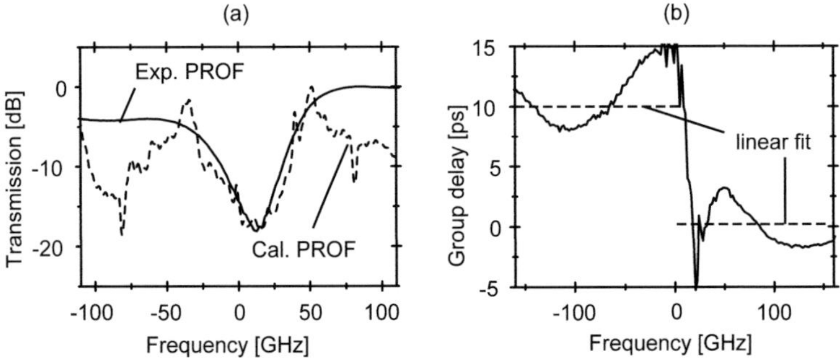

Fig. 4.17. (a) Transmission spectrum of PROF (dashed lines: calculated; solid line: used in experiment). (b) Group delay of the PROF used in experiment. The dashed line is obtained from linear fit.

Complementary Pattern Effects in the XPM-induced Chirp

Now we investigate the complementary pattern effects in the XPM-induced red and blue chirp of the inverted signal. The complementary pattern effects have their origin in the carrier dynamics of the SOA. We therefore used an SOA model from Chapter 2 that encompasses both slow and fast carrier recovery effects. The parameters in Appendix B were adapted for the current SOA and validated experimentally. Two second-order Gaussian filters with a 3dB bandwidth of 120 GHz were used as the RSOF and the BSOF in the simulation to select the spectral components. Simulation results are given in Fig. 4.18. An exemplary input data signal with a bit pattern "0111", Fig. 4.18(a), is launched into the SOA. Fig. 4.18(b) shows the gain evolution with time. The gain decreases with the leading edge of each input pulse and recovers with the falling edge of each pulse. Further, with each sub-sequent bit the SOA gain gradually saturates. Due to Kramers-Kronig relation, a gain change is always accompanied by a phase-shift $\Delta\phi_{NL}$, whose variation per time increment Δt in consequence determines the instantaneous frequency shift Δf (i.e. chirp, see formula of Fig. 4.18(c)). The leading edges of the input pulses induce a red chirp, while the trailing edges induce a blue chirp, Fig. 4.18(c). As we will see below, these two chirps have different pattern dependences.

We first study the pattern dependence of the signal after the RSOF. The first "1" bit in the pulse train induces a strong carrier depletion and a large gain reduction Fig. 4.18(b). A strong carrier-depletion induces a large phase-shift $\Delta\phi_{NL}$ and consequently a large red chirp on the leading edge, Fig. 4.18(c). For subsequent "1" pulses, the SOA does not fully recover. The gain for subsequent "1" pulses is smaller than for the first "1" bit. Therefore the carrier depletion decreases. As the gain reduction due to the carrier depletion decreases, the induced red chirp decreases as well, low part of Fig. 4.18(c). This decreasing red chirp however, by means of the RSOF filter transfer function (see 1,2,3 in Fig. 4.18(d)), and the increasing gain saturation create a bit pattern with decreasing power level, shown in Fig. 4.18(e) bottom.

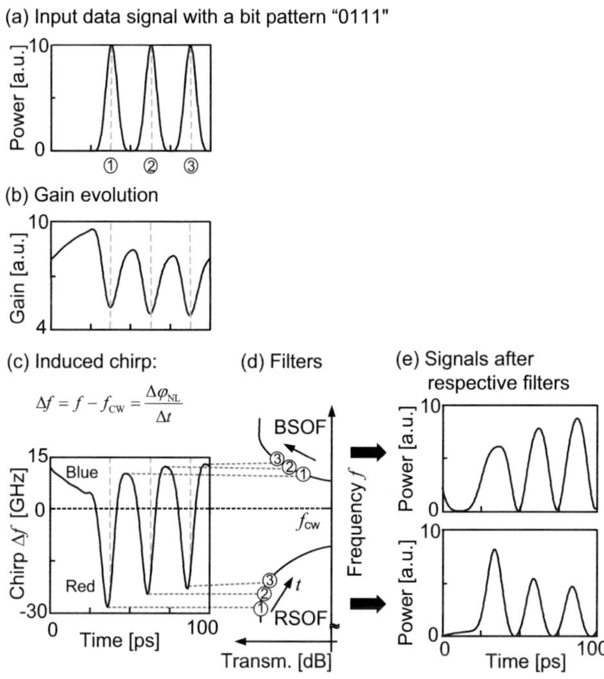

Fig. 4.18. Complementary pattern effects by frequency-amplitude conversion through differently centered filters. (a) Input data pulse train. (b) Simulated gain evolution in SOA and (c) induced frequency chirp in the inverted signal. (d) Schematic filter shape of the BSOF and the RSOF. (e) Pattern dependence in blue and red-spectral components after sending signal (b) with chirp (c) through filter plotted in (d). Dashed lines in (a), (b) and (c) indicate the time positions of the input three pulses.

The bit pattern of the signal behind the BSOF can be understood as follows. The gain recovers with the trailing edge of the first "1" pulse, creating a blue chirp. With subsequent bits launched into the SOA, the SOA is further saturated and the overall gain G decreases with following "1" bits. In turn the carrier depletion due to stimulated emission is now weaker while the current injection is unchanged. As a consequence, the carrier recovery rate is faster for subsequent pulses. This leads to a stronger blue-chirp as also observed in the simulation depicted in Fig. 4.18(c). By positioning the strongest blue-chirped signals closest to the center of the BSOF passband, Fig. 4.18(d), transmission of the strongest chirped pulse is favored. This leads to an increasing power level for the tails of longer patterns, shown in Fig. 4.18(e) top. The coherent superposition of the red- and blue-shifted signal with opposite patterns then leads to the pattern effect compensated output, Fig. 4.16(d). The scheme works well, as long as the frequency-amplitude conversion at the slope of the BSOF can overcompensate the gain saturation. Simulations show that results do not change for a 3 dB filter bandwidth of 80 to 220 GHz.

Large Power Dynamic Range and Wavelength Tolerance

The PROF scheme uses both the red- and blue-shifted spectral components, thus keeps all the information in the spectrum. This advantage leads a to a large input power dynamic range, in which a good signal quality is kept. In our experiments, we obtained a Q^2-factor of 15.8 dB with an input power ($\lambda_{in} = 1530$ nm) of as low as -4.3 dBm while a cw power ($\lambda_{cw} = 1536.1$ nm) is -4 dBm. At an input data power of 12.9 dBm, which is the available maximum in our experiment, and a cw power of 14.7 dBm, we also got a Q^2-factor of 18.2 dB. In fact, throughout this input power dynamic range of at least 17.2 dB with an adjustment of the cw power, we always got Q^2-factors above 15.6 dB, shown in Fig. 4.19. Note that the cw power adjustment for respective input data powers was stopped as long as we have a Q^2-factor above 15.6 dB. All of Q^2 values were measured with a bit-error rate test (BERT) and then interpolated from the inverse complementary error function. Yet, as it is not clear on what noise distribution we have after an all-optical wavelength conversion operation. The Q^2 values may differ based on the method used. However, the quantitative values given here are still indicative for the quality of the signal.

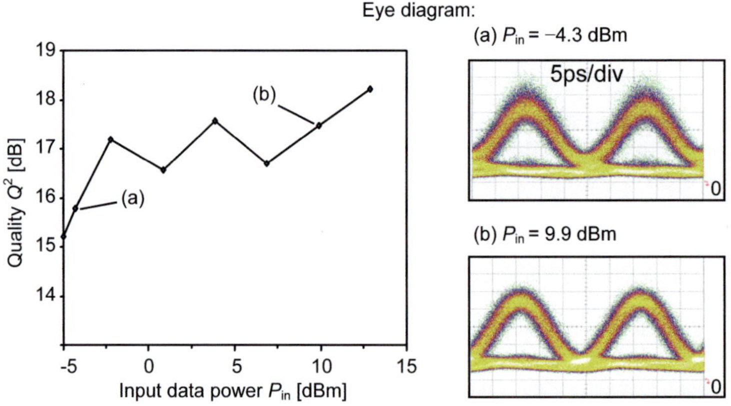

Fig. 4.19. Signal qualities of the converted signal for various input data powers without adaptation of filter parameters but with adaptation of the cw input power levels, while $\lambda_{in} = 1530$ nm and $\lambda_{cw} = 1536.1$ nm. Eye diagram for $P_{in} = -4.3$ dBm and 9.9 dBm are given in (a) and (b) respectively.

Except a large input power dynamic range, the PROF scheme features a large tolerance of the operating wavelength. In the experiment, we kept the input power $P_{in} = 12.9$ dBm at $\lambda_{in} = 1530$ nm and the cw power $P_{cw} = 14.7$ dBm, while the operating wavelength of the cw signal λ_{cw} varied from 1536.0 nm to 1536.4 nm. The measured Q^2-factor is depicted in left part of

Fig. 4.20. We see that a wavelength tolerance for a Q^2-factor above 15.6 dB is about 0.29 nm. Eye diagram for λ_{cw} = 1536.05 nm, 1536.15 nm and 1536.3 nm are given in Fig. 4.20(a), (b) and (c) respectively. All the eyes are clean and open.

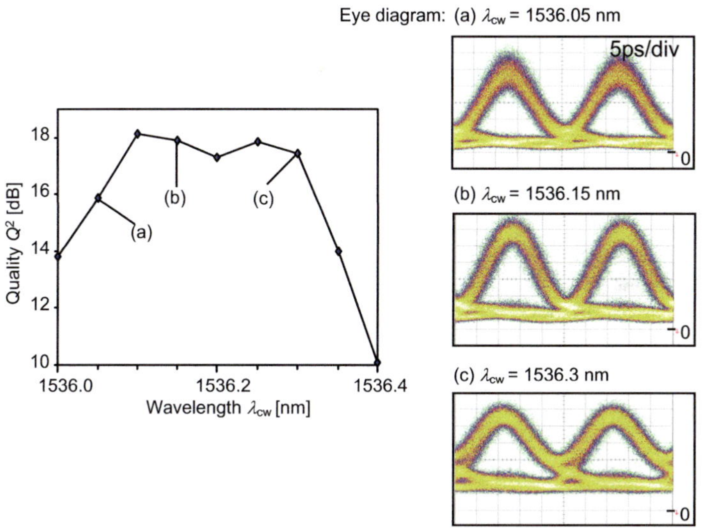

Fig. 4.20. Signal qualities of the converted signal for various cw wavelengths without adaptation of filter parameters, while the cw power P_{cw} = 14.7 dBm and the input power P_{in} = 12.9 dBm at λ_{in} = 1530 nm. Eye diagrams for λ_{cw} = 1536.05 dBm, 1536.15 dBm and 1536.3 nm are given in (a), (b) and (c) respectively.

We now look more closely at the converted signals in this 0.29 nm operation range. Exemplary patterns and spectra of the converted signals at λ_{cw} = 1536.05 nm, 1536.15 nm and 1536.3 nm are given in the right and left columns of Fig. 4.21(a), (b) and (c). At λ_{cw} = 1536.15 nm, Fig. 4.21(b), the pattern effect is almost totally suppressed, while the spectrum is more or less alike an on-off keying (OOK) signal. At λ_{cw} = 1536.05 nm in Fig. 4.21(a), there is still a residual pattern effect, especially at the first bit "1" after some "0"s. At λ_{cw} = 1536.3 nm in Fig. 4.21(c), the pattern effect is also not observable. However, it is obtained at the expense of part of the spectrum. We can see in Fig. 4.21(c) that the blue part of the signal spectrum is suppressed. This leads to a broadening of the output pulses and a limited transmission distance, if no other remedy is applied.

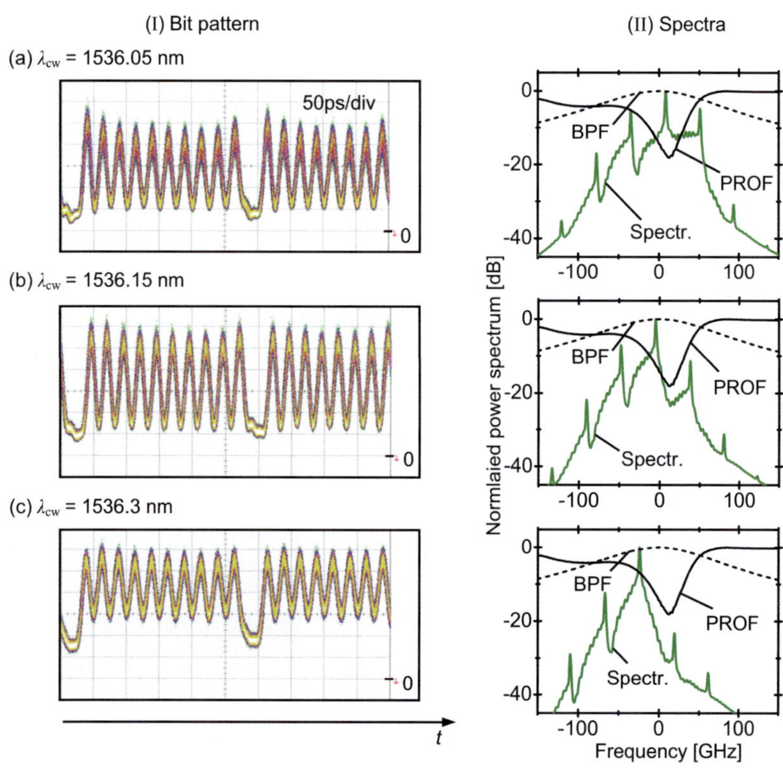

Fig. 4.21. Bit pattern and spectra of wavelength converted signals for different center wavelengths λ_{cw} after a fixed pulse reformatting optical filter (PROF). Note that the spectra are plotted on the demodulated frequency axis, with a modulation wavelength 1536.1 nm. The dashed lines in right figures are one band-pass filter (BPF) with a 1 nm FWHM-bandwidth.

Last but not least, we look at how the mode beating effect influences the combined signals. With respect to the red-shifted signals, Fig. 4.22(a), the blue-shifted signals are delayed from −20 ps to 20 ps, Fig. 4.22(b). Note that all of the time delays are the relative values to the optimum time delay for the results shown in Fig. 4.16. The combined signal, Fig. 4.22(c), still shows a Q^2-factor of 15.6 dB for a relative time delay of − 10 ps. This is the case, where there is almost no absolute time delay between the red- and blue-shifted signals, since the optimum time delay for Fig. 4.16 is ∼10 ps. However, the pulse width in this case is wider. This observation confirms the discussion in section 4.2.1.

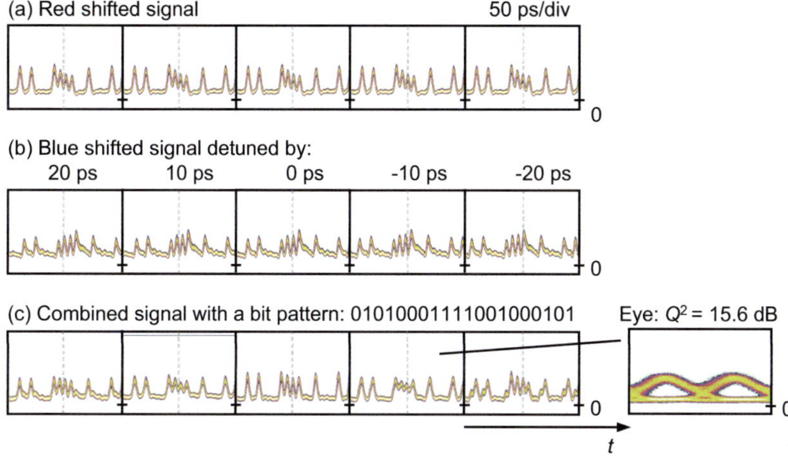

Fig. 4.22. Results from (a) red-shifted signals and (b) blue-shifted signals with different time delays in-between to (c) combined signals, where all of the time delays are the deviations from the optimum time delay.

4.3 Pattern Effect Mitigation Technique Using Red or Blue-Shift Optical Filters

The PROF scheme exploits the fast chirp effects in the converted signal after the SOA and uses both the red- and blue-shifted spectral components, while schemes with a single red- or blue-shift filter [44], [49] and [62] reject part of the spectrum. This leads to a low power level of the filtered signal, which limits the signal quality if a receiver is applied. From an information technological point of view, rejection of spectral components with information should be also avoided.

Nevertheless, the schemes with single blue-shift or red-shift filter are attractive due to their demonstrations at a transmission speed as high as 320 Gbit / s [50] and their long-haul transmission over 42 Nodes and 16,800 km [44]. Indeed, as pointed out before, the opposite pattern effect between blue chirp and gain saturation can be utilized as pattern mitigation technique as well [49] and [62]. However, so far it is not clear, how a red-shift filter suppresses the pattern effect in SOA-based wavelength conversion.

In the following, we will first discuss the schemes using a blue- or red-shift optical filter. Then we will compare the results by using three types of filters.

4.3.1 Scheme with a Blue or Red Shifted Optical Filter

Working Principle

We first consider the wavelength conversion assisted by a blue-shift optical filter (BSOF). We show that by detuning the BSOF to different positions inverted or non-inverted converted signals are achievable. The working principle is explained in Fig. 4.23.

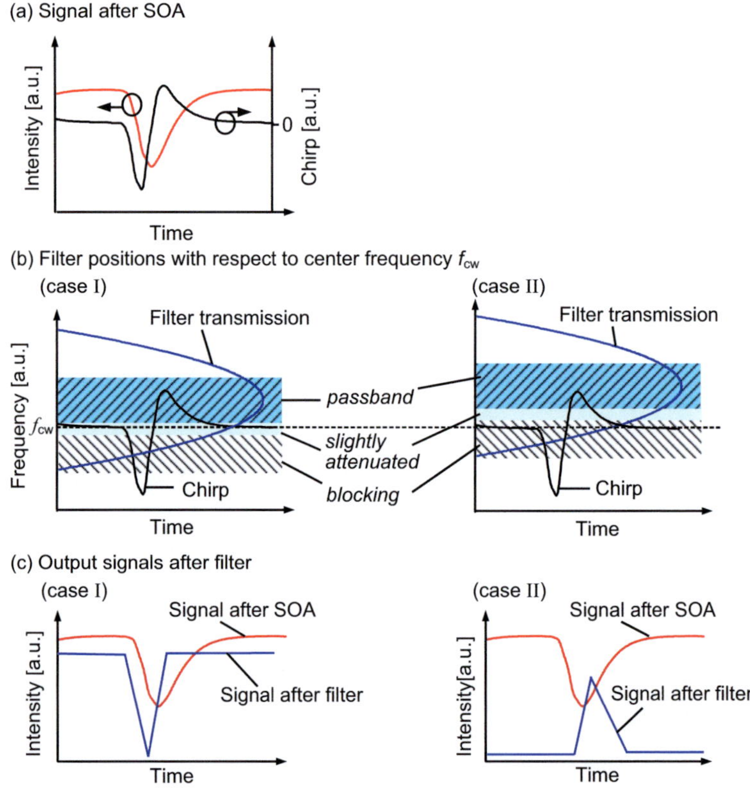

Fig. 4.23. Working principle of using differently centered blue-shift optical filters (BSOFs) to re-shape (a) an inverted pulse after SOA into (c, case I) an improved inverted pulse and (c, case II) a non-inverted pulse. Seen in (b, case I), the blue-chirped component passes through the filter, the optical carrier is slightly attenuated and the red-chirped component is mostly blocked. Seen in (b, case II), the blue-chirped component passes through the filter, while other spectral components are blocked.

Fig. 4.23(a) illustrates the intensity and the frequency chirp of the output signal after an SOA. As the BSOF is detuned to the position (I), left figure in Fig. 4.23(b), the blue-shifted spectral component passes the filter, while the red-shifted spectral component is mostly blocked and the center spectral component is attenuated slightly. Therefore the signal at different times sees different transmission of the filter. In the carrier depletion regime, the signal experiences a red chirp, Fig. 4.23(a). The broadened signal spectrum is now out of the passband of the filter, thus the signal transmission is highly suppressed. In the carrier recovery regime, a blue chirp is imposed onto the signal. So the signal passes. In an ideal case, the un-modulated signal (not-chirped spectral component) is attenuated so that the output intensity after the filter is equalized to the blue-chirped one. The output signal, as case (I) in Fig. 4.23 (c), is then an inverted signal but with an improved extinction ratio.

Positioning of the BSOF into higher frequency range can also make a non-inverted output signal be possible. In the case (II) in Fig. 4.23(b), both center and red-chirped spectral components are blocked by the BSOF. Only the blue-chirped spectral component can pass. As a result, the output signal after the BSOF has a non-inverted waveform, case (II) in Fig. 4.23(c).

The working principle of using a red-shift optical filter (RSOF) is same to that of using a BSOF, Fig. 4.23, except the center frequency of the RSOF is now lower than the cw frequency.

Experimental Results

The discussion above was confirmed with experiments. The BSOF experiment setup is almost same to the PROF scheme in Fig. 4.15, except the upper RSOF part is blocked. The two 50:50 couplers were kept, in order to compare results by using the BSOF, the RSOF and the PROF. In Fig. 4.24, we plotted the wavelength conversion results after differently positioned BSOFs. The wavelength offset of the BSOF $\lambda'_{0,B} - \lambda_{0,B}$ varies from -0.3 nm to 0.3 nm, (a)-(g) in Fig. 4.24(I). Note that the BSOF center wavelength $\lambda_{0,B}$, i.e. $\lambda'_{0,B} - \lambda_{0,B} = 0$, was chosen for a best non-inverted output signal. Fig. 4.24(II) and (III) are evolution of the waveform polarities and eye diagrams of the converted signals. As expected, a good inverted signal was also got at $\lambda'_{0,B} - \lambda_{0,B} = 0.3$ nm, at which the center spectral component passes through the BSOF. Note that detuning $\lambda'_{0,B}$ was realized by detuning the cw center wavelength λ_{cw}.

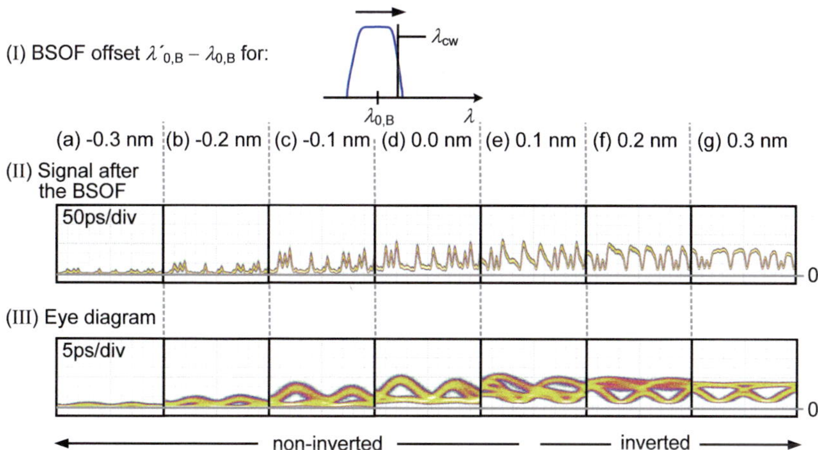

Fig. 4.24. (I) Wavelength offset of the blue-shift optical filter (BSOF), $\lambda'_{0,B} - \lambda_{0,B}$, from -0.3 nm to 0.3 nm. (II) and (III) are evolution of the waveform polarities and eye diagrams of the converted signal.

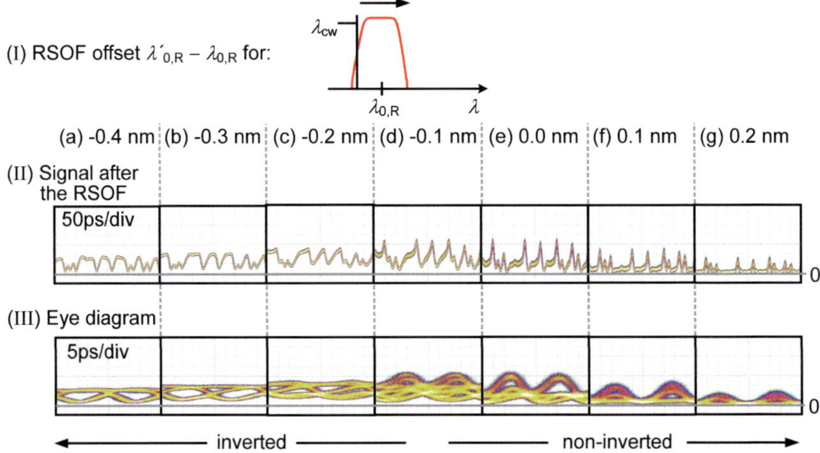

Fig. 4.25. (I) Wavelength offset of the red-shift optical filter (RSOF), $\lambda'_{0,R} - \lambda_{0,R}$, from -0.4 nm to 0.2 nm. (II) and (III) are evolution of the waveform polarity and eye diagrams of the converted signal.

We now discuss the evolution of the signal waveforms after a RSOF. The experiment setup was changed by blocking the lower BSOF part in Fig. 4.15. Either inverted or non-inverted converted signal were also got in the experiment, shown in Fig. 4.25. The wavelength offset of the RSOF $\lambda'_{0,R} - \lambda_{0,R}$ varies from -0.4 nm to 0.2 nm, (a)-(g) in Fig. 4.25(I). Fig. 4.25(II) and (III) are evolution of the waveform polarity and eye diagrams of the converted signal. The offset $\lambda'_{0,B} - \lambda_{0,B} = 0$ is the position, where a good non-inverted signal was got. A good inverted signal was then got as $\lambda'_{0,B} - \lambda_{0,B} = -0.40$ nm, which means that the cw spectral component moves into the RSOF passband.

4.3.2 Comparison of Pattern Effect Mitigation Techniques Using Optical Filters

Pattern effect mitigation using a blue-shift optical filter

We now focus on the quality of the non-inverted signal after the filter. In the BSOF experiment, we varied the cw center wavelength λ_{cw}, effectively detuning the BSOF. The Q^2-factors were only measurable as λ_{cw} varies between 1536.0 nm and 1536.15 nm, and are compared in Fig. 4.26. Eye diagrams for $\lambda_{cw} = 1535.95$ nm, 1536.05 nm, and 1536.15 nm are given in Fig. 4.26(a), (b) and (c). The corresponding patterns and spectra of the converted signals are given in the right and left columns of Fig. 4.27(a), (b) and (c). The power transmissions of the fixed BSOF and the 1 nm wide bandpass filter (BPF) centered at $\lambda_{cw} = 1536.1$ nm are also plotted in Fig. 4.27. Note that the spectra are plotted on the baseband frequency axis, with a modulation wavelength 1536.1 nm.

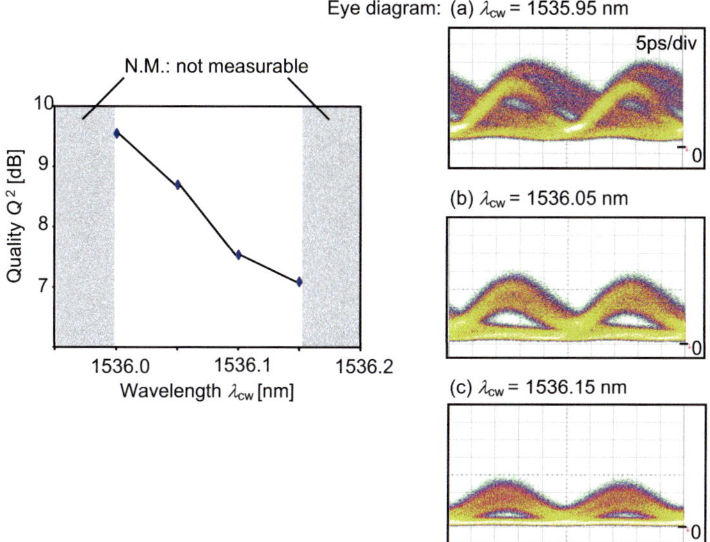

Fig. 4.26. Signal qualities of the blue-shifted signal for various cw wavelengths without adaptation of filter parameters, while the cw power P_{cw}=14.7 dBm and the input power P_{in}=12.9 dBm for λ_{in}= 1530 nm. Eye diagrams for λ_{cw}= 1535.95 nm, 1536.05 nm and 1536.15 nm are given in (a), (b) and (c) respectively.

We see that the largest Q^2-factor in Fig. 4.26 is only 9.5 dB. This is mostly because the blue-chirped spectral component is usually weak, seen in right figure of Fig. 4.16(a). This limits the extinction ratio of the non-inverted converted signal after the BSOF. We also see that the pattern effect is more or less mitigated for the decreasing λ_{cw}, Fig. 4.27(I,c)-(I,a). However, as the converted signal moves into the passband of the BSOF, the data polarity transforms from the non-inverted to the inverted format. At a wavelength λ_{cw} = 1536.05 nm and λ_{cw} = 1536.15 nm in Fig. 4.27(I,b) and (I,c), the frequency-amplitude conversion at the slope of the BSOF overcompensated the gain saturation effect. This is exactly the effect which we utilized in the PROF scheme. For the BSOF scheme, the overcompensated pattern effect tells that the BSOF slope close to the cw wavelength is too steep. So, if a less steep BSOF is used, the frequency-amplitude conversion at the slope of the BSOF can compensate the gain saturation effect. The pattern effect is then mitigated after the BSOF.

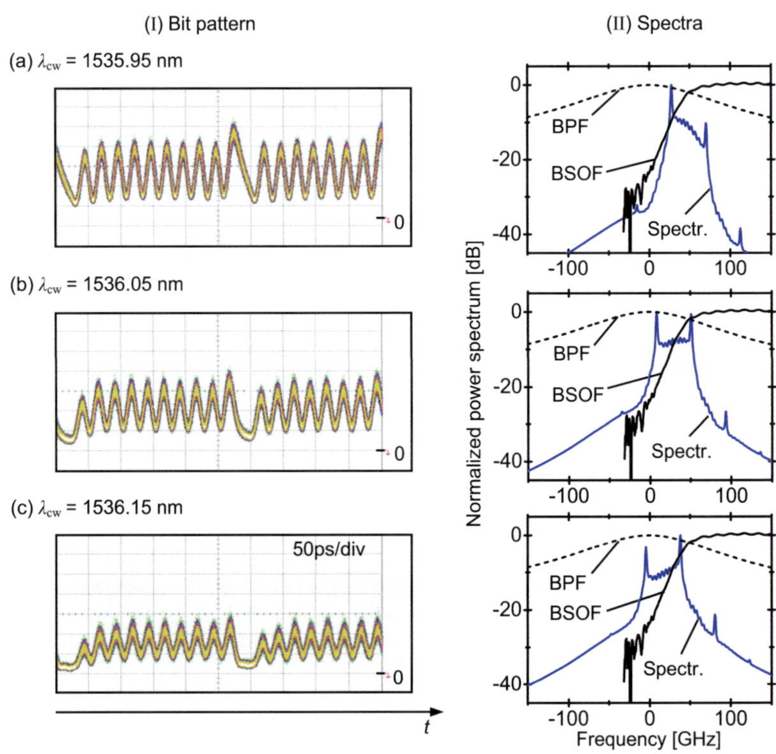

Fig. 4.27. Pattern effects and spectra of the blue-shifted signal with different center wavelengths λ_{cw} after a fixed red shift optical filter (BSOF) and a 1 nm wide bandpass filter (BPF) centered at λ_{cw} = 1536.1 nm. The pattern effect is mitigated for the decreasing λ_{cw}, as the converted signal moving into the passband of the BSOF, but the data polarity transforms from the non-inverted to the inverted format. Note that the spectra are plotted on the demodulated frequency axis, with a modulation wavelength 1536.1 nm.

Pattern effect mitigation using a red-shift optical filter

We now consider the wavelength conversion assisted by a RSOF. By varying the cw center wavelength λ_{cw} between 1536.0 nm and 1536.4 nm, which effectively detunes the BSOF, we measured the Q^2-factors and compare them in Fig. 4.28. We can see a wavelength tolerance of ~ 0.2 nm for a Q^2-factor above 15.6 dB. Eye diagrams for λ_{cw} = 1536.05 nm, 1536.3 nm, and 1536.4 nm are given in Fig. 4.28(a), (b) and (c). The corresponding exemplary patterns and spectra of the converted signals are given in the right and left columns of Fig. 4.29(a), (b) and (c). The power transmissions of the fixed BSOF and the 1 nm wide bandpass filter (BPF) centered at λ_{cw} = 1536.1 nm are also plotted in Fig. 4.29. Note that the spectra are plotted on the demodulated frequency axis, with a modulation wavelength 1536.1 nm.

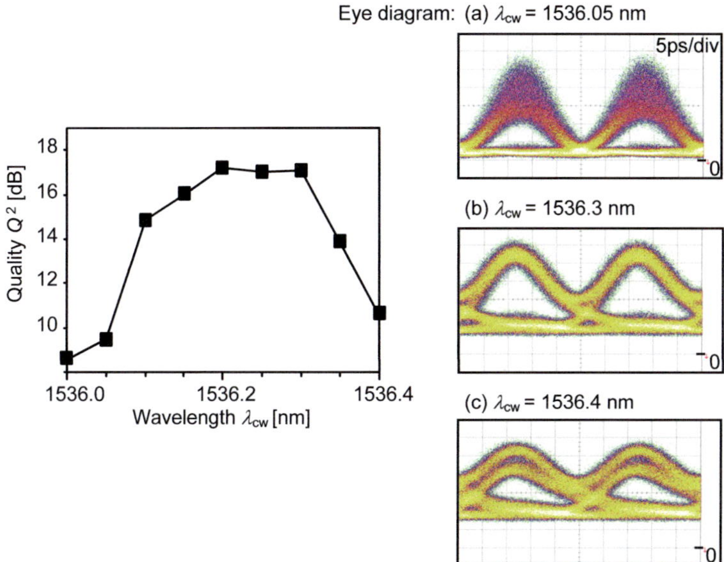

Fig. 4.28. Signal qualities of the red-shifted signal for various cw wavelengths without adaptation of filter parameters, while the cw power P_{cw} = 14.7 dBm and the input power P_{in} = 12.9 dBm for λ_{in} = 1530 nm. Eye diagrams for λ_{cw} = 1536.0 nm, 1536.1 nm and 1536.3 nm are given in (a), (b) and (c) respectively.

While the RSOF is mainly used to suppress the center spectral component, to have a non-inverted waveform, the pattern effect compensation at λ_{cw} = 1536.3 nm is achieved by utilizing the 1 nm wide BPF. From the filter transmission spectra (dashed lines) shown in Fig. 4.29(II), the strongest red-chirped signals always experience the lowest transmission. As shown in Fig. 4.18(c), the strongest chirp happens in the leading pulse of a long pattern. Such a long pattern also experiences a pattern effect in the gain, decreasing over the time shown in Fig. 4.18(b). The superposition of opposite pattern effect between the red-chirp induced frequency-amplitude conversion at the slope of the BPF and the gain saturation then leads to the pattern effect compensated output, as λ_{cw} = 1536.3 nm in Fig. 4.29(b), and also an open eye, Fig. 4.28(b). This leads to a Q^2-factor of 17.1 dB, as given in left part of Fig. 4.28.

Now, the frequency-amplitude conversion at the slope of the BPF can be clearly compared by varying the position of the red chirp on the filter slope. As shown in Fig. 4.29(a), the red chirps still locate on the relative flat slope. This leads a weak frequency-amplitude conversion due the red chirp. The output signal still shows a pattern effect due to the gain. On the other hand, if the red chirp is on the more steep slope of the filter, the output signal is even over-compensated, Fig. 4.29(c).

Thus, a practical RSOF reshaping the converted signal and compensating the pattern effect simultaneously should comprise two optical filters. The first filter should suppress the center spectral component, to get a non-inverted waveform. The second filter should be positioned such that the strongest red chirp experiences the minimum transmission, to induces an opposite pattern effect to the gain saturation. As a result, a non-inverted signal appears after the filter, and the pattern effect can also be mitigated.

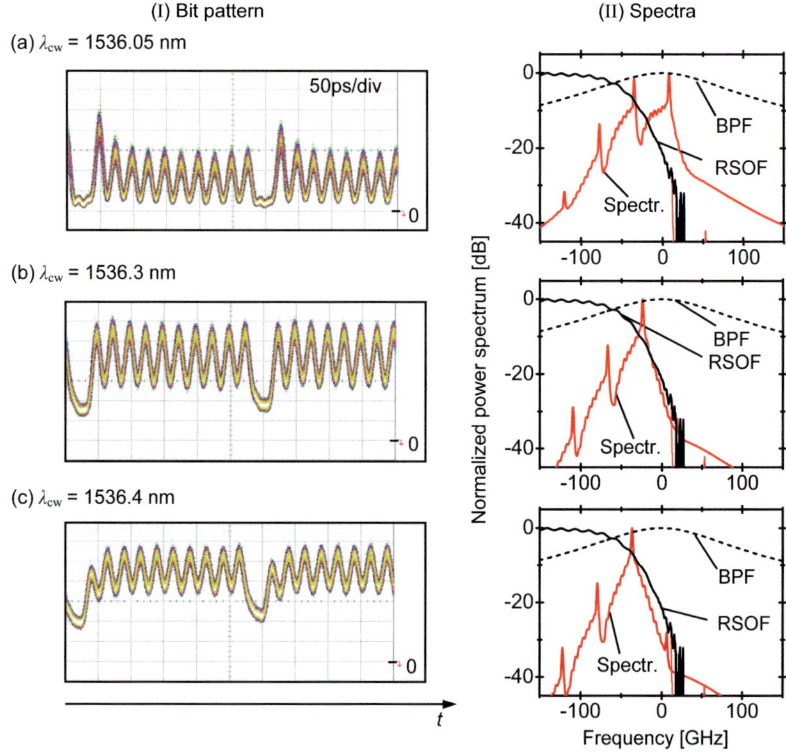

Fig. 4.29. **Pattern effects and spectra of the red-shifted signal with different center wavelengths λ_{cw} after a fixed red shift optical filter (RSOF) and a 1 nm wide bandpass filter (BPF) centered at λ_{cw} = 1536.1 nm. The pattern effect is compensated at (b) λ_{cw} = 1536.3 nm, while not compensated at (a) λ_{cw} = 1536.1 nm and over-compensated at (c) λ_{cw} = 1536.4 nm. In addition, the DC component in the converted signals increases for the increasing λ_{cw}, as the converted signal moves into the passband of the RSOF. Note that the spectra are plotted on the demodulated frequency axis, with a modulation wavelength 1536.1 nm.**

Comparison to the scheme using a pulse reformatting optical filter

In summary, the quality factors of the red-shifted (RS) signals after a RSOF, the blue-shifted (BS) signals after a BSOF and the combined signals after a PROF are compared in Fig. 4.30(I). Fig. 4.30(II) shows the signal spectra for λ_{cw} = 1536.05 nm, 1536.15 nm and 1536.3 nm. We see that a good signal with a center wavelength above 1536.2 nm is mainly contributed by the red-shifted signal. This is also confirmed by the signal spectrum for λ_{cw} = 1536.3 nm, shown in Fig. 4.30(II,c). As λ_{cw} goes down, especially as λ_{cw} = 1536.05 nm, the

Q^2-factor of the red-shifted signal decreases below 10 dB. However, the combined signal still has a Q^2-factor of ~ 16 dB. This is thanks to the pattern compensation technique utilizing the complementary pattern effects between the red and blue chirp. This technique is realized by using a PROF and a bandpass filter. This scheme gives a wavelength tolerance of about 0.29 nm for a Q^2-factor above 15.6 dB. Note that the blue-shifted signals got in our experiment were not good. The measurable Q^2-factors for λ_{cw} from 1536.0 nm to 1536.15 nm were below 10 dB. This is due to the weak blue-chirped component got from the XPM in the SOA. If a strong blue chirp is got and a BSOF is carefully selected as in [49], we expect a larger wavelength tolerance by combining the red-shifted and blue-shifted signal in the PROF scheme.

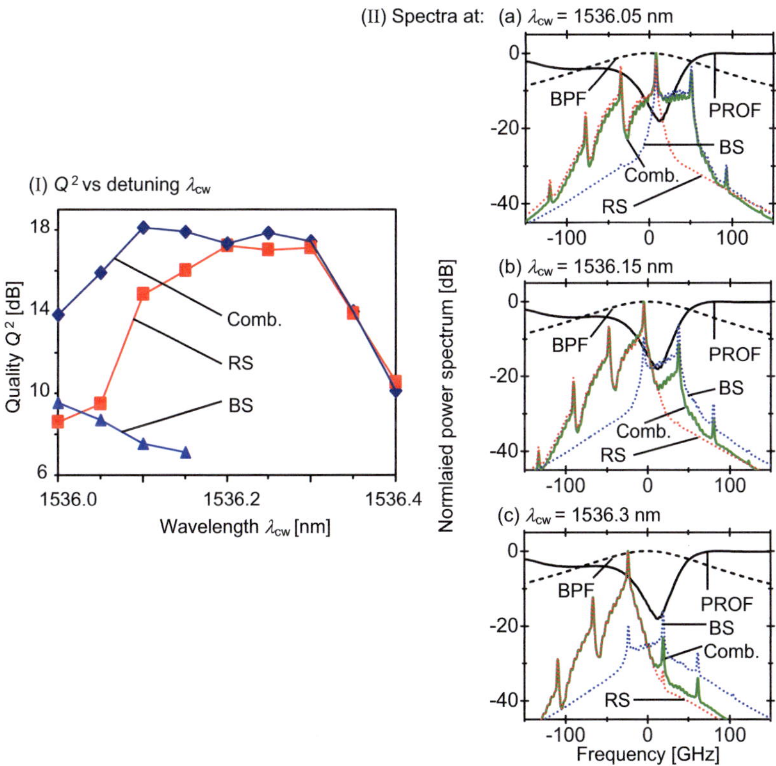

Fig. 4.30. Signal qualities of the converted signal for various cw wavelengths without adaptation of filter parameters, while the cw power P_{cw} = 14.7 dBm and the input power P_{in} = 12.9 dBm at λ_{in} = 1530 nm. Spectra of combined signals for λ_{cw} = 1536.05 nm, 1536.15 nm and 1536.3 nm are given in (a), (b) and (c) respectively. In comparison, the spectra of blue-shifted part (BSP) and red-shifted part (RSP) before they are combined are also plotted in the respective figures. Note that the spectra are plotted on the demodulated frequency axis, with a modulation wavelength 1536.1 nm.

5 Wavelength Conversion Using an SOA-based Sagnac Loop

All-optical wavelength converters based on the Sagnac interferometer configuration with a semiconductor optical amplifier (SOA) are very well suited for non-return-to-zero (NRZ) wavelength conversion [19], [30] and [87]. The NRZ modulation format gains interests due to the simplicity of this transmission format, its reasonable dispersion tolerance and spectral efficiency. As an interferometric configuration, the Sagnac loop can output signal with a good extinction ratio, [12] and [31].

The SOA-related pattern effect can also be mitigated by using the Sagnac loop. The idea behind is simple [19]: The SOA's output power and phase are proportional to a moving average (i.e. an integral) over the input power. The SOA is integrated in a Sagnac interferometer. If the phase relation in the loop is correctly set, the Sagnac interferometer is to differentiate the SOA's output and to restore the shape of the incident signal. This working principle is similar to the usage of a delay interferometer (DI) after the SOA in Chapter 4, where the low-pass characteristic of the SOA is equalized by the high-pass characteristics of the DI.

Operating with the SOA in the Sagnac loop also offers one advantage: The SOA can operate at a relatively slow speed, and since the output is proportional to the integral over the input signal high frequencies, noise is washed out. This integration over the input signal leads a signal regeneration. The expression of the output with an integral over the high-frequency component of the input signal has been given in [19] and [31], and is outlined in Appendix E.

The operation principle of the NRZ wavelength conversion using the Sagnac loop has been discussed in [19] and [88]. Yet, good performance is only obtained if an SOA with proper physical characteristics, such as the XGM and XPM, is selected. In those references [19] and [88], the influence of the SOA properties, especially the SOA carrier recovery time, on the performance of the wavelength conversion has neither been discussed nor been well understood.

In this Chapter, we show how operation speed and SOA dynamics are related. Most surprisingly we find in the simulation that SOA recovery times should neither be too fast nor be too slow for a particular operation speed. More specifically, it is shown that a carrier recovery time between 2 and 3 times of one bit duration will give an optimum output NRZ signal. Further, the influence of the SOA length on the performance is discussed. We find that short SOAs are preferred to be used in the Sagnac loop.

5.1 Theory

In this section, we introduce the configuration and operation principle of the wavelength conversion using the Sagnac loop. We start from a Sagnac interferometer working as a nonlinear fiber-loop mirror. The introduction is then extended to the wavelength conversion using a Sagnac loop.

5.1.1 Configuration

Fiber-loop mirror

Fig. 5.1 shows schematically how a fiber coupler can be used to make a Sagnac interferometer. It is made by connecting a piece of long fiber to the two output ports of a fiber coupler to form a loop. The input optical signal is spit into two counterpropagating parts that propagate through the same optical path and interfere at the coupler coherently. These two counterpropagating parts are also termed as a clockwise propagating signal and a counterclockwise propagating signal. The relative phase difference between these two counterpropagating signals determines whether an input signal is reflected or transmitted by the Sagnac interferometer. Ideally, if a 3-dB fiber coupler is used, any input is totally reflected. In this case, the Sagnac loop acts as a perfect mirror. For this reason, such a device is called as a fiber-loop mirror.

A Sagnac loop can be also used as a fiber-loop reflector, if a birefringent element is integrated in the loop. The birefringent element introduces a phase difference between two counterpropagating signals. Thus, the input signal passing through the Sagnac loop can be guided to the transmission port or to the reflection port in Fig. 5.1, depending on the choice of the birefringent element. The configuration and principle are given in Appendix D.

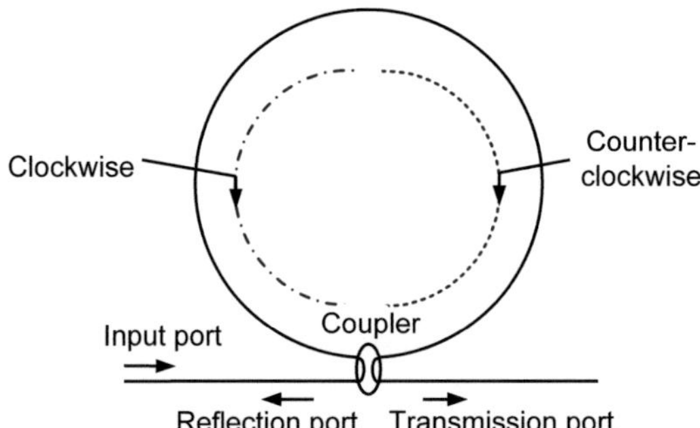

Fig. 5.1. Scheme of a Sagnac interferometer working as a nonlinear fiber-loop mirror.

A Sagnac interferometer can also be used to realize all-optical wavelength conversion, if a nonlinear element such as an SOA is integrated in the Sagnac interferometer. A probe signal launched in the input port is split into counterpropagating parts passing through the nonlinear element. In this nonlinear element, either cross-gain or cross-phase modulation effects can be exploited. These nonlinear effects are induced by another input signal, also called as control signal. The transmission of the probe signal from the input port to the transmission port is then controlled by the control signal. In other words, the wavelength conversion is performed.

Wavelength converter using a Sagnac loop

The wavelength conversion scheme using a Sagnac loop is shown in Fig. 5.2. An SOA is asymmetrically arranged in the Sagnac loop, offset by a time delay $\Delta T/2$ away from the center of the Sagnac loop. A coupler C1 splits a continuous wave (CW) signal with wavelength

λ_1 into two counter-propagating CW waves. They are termed as clockwise propagating wave P_{cw}^c and counter-clockwise propagating wave P_{cw}^{cc}, with superscripts "c" and "cc"[14], respectively. Another coupler C2 introduces an input data signal at a wavelength λ_2 into the Sagnac loop. This input signal induces nonlinear amplitude and phase changes on the two counter-propagating waves through the SOA. The two counter-propagating waves combine in the coupler C1 and form a wavelength converted signal P_{out}. A bandpass filter centered at λ_1 is then used at the output to block the transmitted input data signal. A polarization controller PC1 is used to control the polarization of the CW signal. Another polarization controller PC2 is connected on the right-hand side of the Sagnac loop and close to the coupler C1.

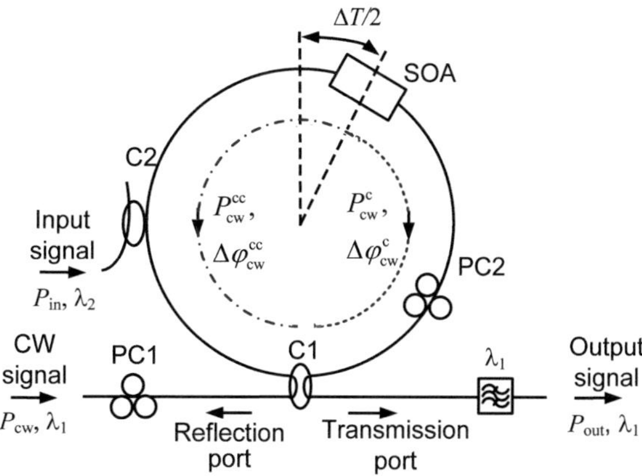

Fig. 5.2. All-optical wavelength conversion scheme by applying an SOA-based Sagnac loop. An SOA is asymmetrically located in the Sagnac loop, offset by a time delay of $\Delta T/2$. PC1 and PC2 are polarization controllers, and C1 and C2 are couplers. By exploiting the cross-gain and cross-phase modulation effects, an input signal P_{in} at the wavelength λ_2 is converted to an output signal P_{out} at a new wavelength λ_1. A bandpass filter centered at $\Delta T/2$ is used to block the input signal. $P_{cw}^{c(cc)}$ and $\Delta\varphi_{cw}^{c(cc)}$ are the power and phase of the CW signals inside of the Sagnac loop, where the superscripts "c" and "cc" denote the clockwise and counter-clockwise propagating cw signal respectively.

To understand the operation principle of this type of wavelength converter, the polarization controller PC2 needs more explanations. The polarization controller PC2 is a birefringent device, which adds a phase difference ψ between the two counter-propagating waves, [19] and [88]. The polarization controller PC2 works similar to the phase shifter used in the Mach-Zehnder interferometer [38] or in the delay interferometer [42]. It is often that the polarization controller PC2 is a waveplate[15], which has a retardation ϕ and an orientation θ. In the Carte-

---

[14]Note that the superscript "cc" is not the term *c.c.* used in Eq. (2.9), which indicates the complex conjugate field component.

[15]Waveplates – also known as retardation plates – are optical elements that create a phase shift in the transmitted light with the help of a birefringent crystal quartz.

sian coordinate system, the orientation θ is the angle of the fast axis e with respect to the y-polarization direction of the complex field, seen in Fig. 5.3. The retardation ϕ is the phase lead of the field component on the fast axis e with respect to the field component on the slow axis o of the birefringent medium (in a so-called half waveplate, the phase lead is 180°).

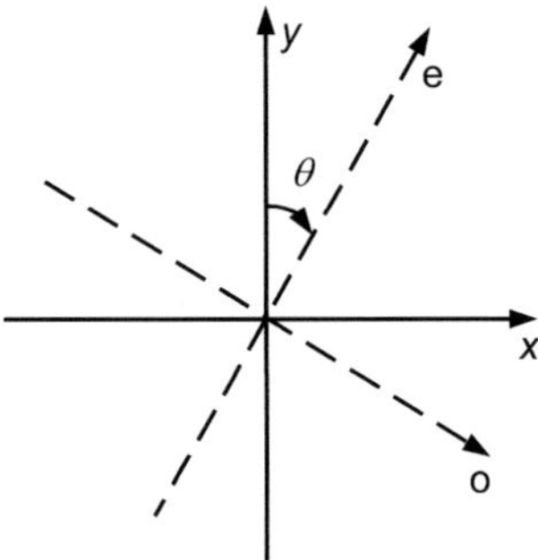

Fig. 5.3. The orientation θ of the fast axis e with respect to the y-axis.

Such a waveplate can be described by the Jones matrix $\underline{J}$ [61] and [88],

$$\underline{J} = \begin{pmatrix} J_{xx} & J_{xy} \\ J_{yx} & J_{yy} \end{pmatrix}, \tag{5.1}$$

where the elements of the Jones matrix are[16]:

$$J_{xx} = \exp(j\phi)\sin^2\theta + \cos^2\theta, \tag{5.2}$$

$$J_{xy} = J_{yx} = [\exp(j\phi) - 1]\cos\theta\sin\theta, \tag{5.3}$$

$$J_{yy} = \exp(j\phi)\cos^2\theta + \sin^2\theta. \tag{5.4}$$

Here the polarization dependent losses are neglected. Details of Jones matrix for a waveplate can be found in Appendix D.

5.1.2 Wavelength Conversion for Non-Return-to-Zero Modulation Signal

In the section, we explain how an SOA-based Sagnac loop performs wavelength conversion for a non-return-to-zero (NRZ) modulation signal. The optical field at an angular frequency ω_0, or at the wavelength λ, is denoted by $E(z,t) = A(z,t)\exp[j(\omega_0 t - \beta_0 z)]$, where $\beta_0 = 2\pi n_r/\lambda$ is the real propagation constant and n_r is the refractive index. $A(z,t) = \sqrt{P(z,t)}\exp[j\Delta\varphi(z,t)]$ is the slowly-varying envelope associated with the optical power $P(z,t)$ and the optical phase shift $\Delta\varphi(z,t)$. By assuming the optical phase in the station-

---

[16]Neglecting the polarization dependent losses in the waveplate, the elements of a Jones matrix fulfill the unitary condition $|J_{xx}|^2 + |J_{yx}|^2 = 1, |J_{xy}|^2 + |J_{yy}|^2 = 1, J_{xx}J_{xy}^* + J_{yx}J_{yy}^* = 0$.

ary state is zero, $\Delta\varphi(z,t)$ is then the nonlinear phase shift picked up by the signal when propagating to the position z inside the SOA. For an SOA with a length of L, the phase shift $\Delta\varphi$ at the output $z = L$ is thereby related to the refractive index change Δn_r (as a function of the carrier density N) by $\Delta\varphi(t) = (-2\pi/\lambda)\int_0^L \Delta n_r\left(N(z,t)\right)\mathrm{d}z$. Similarly, the single pass gain[17] $G(t)$ of an SOA is related to the net gain g by $G(t) = \exp(\int_0^L g\left(N(z,t)\right)\mathrm{d}z)$. Note that the change of the refractive index Δn_r and the gain G are dependent on the input power. The power gains for the clockwise and counter-clockwise propagating CW signals are G_{cw}^c and G_{cw}^{cc}, and the phase shifts of these two signals after the SOA are $\Delta\varphi_{cw}^c$ and $\Delta\varphi_{cw}^{cc}$, see Fig. 5.2. As derived in [88], the intensity transmission T of the CW signal from the input port to the transmission port of the coupler C1, Fig. 5.2, is given as

$$T = \frac{1}{4}\left[G_{cw}^c + G_{cw}^{cc} - 2\gamma\sqrt{G_{cw}^c G_{cw}^{cc}}\,\cos\left(\Delta\varphi_{cw}^{cc} - \Delta\varphi_{cw}^c + \psi\right)\right], \tag{5.5}$$

where

$$\gamma = \sqrt{[\cos^2(2\theta) + \cos\phi\sin^2(2\theta)]^2 + [\sin\phi\sin(2\alpha)\sin(2\theta)]^2}, \tag{5.6}$$

$$\psi = \tan^{-1}\left(\frac{\sin\phi\sin(2\alpha)\sin(2\theta)}{\cos^2(2\theta) + \cos\phi\sin^2(2\theta)}\right). \tag{5.7}$$

In Eq. (5.6) and (5.7) the coefficient α is the absorption of the fiber link in the Sagnac loop. Note that the coupling loss through the coupler C2 can be included in the fiber link loss. As seen in Eq. (5.5), ψ is the phase offset for the transmission. By carefully adjusting retardation ϕ and orientation θ, one can control this phase offset [55] and also the relative phase between two counter-propagating signals.

With Eq. (5.5)-(5.7), we now can discuss the operation principle of the wavelength conversion for NRZ signals, shown in Fig. 5.4. The input data signal carrying information, Fig. 5.4(a), induces nonlinear changes of the gain and the refractive index in the SOA. Both clockwise and counter-clockwise propagating signals, due to cross-gain modulation (XGM) and cross-phase modulation (XPM), pick up the nonlinear change of the gain and also experience the phase shift. The powers P_{cw}^c and P_{cw}^{cc} and the phase shifts $\Delta\varphi_{cw}^c$ and $\Delta\varphi_{cw}^{cc}$ of two counterpropagating signals entering the coupler C1 are shown as dotted and dash-dotted lines in Fig. 5.4(b) and (c), respectively. We assume that the gain/phase fully reduce (with a hypothetical 0 output power after a full gain reduction) or recover with the same time constant, e.g. 2 times of one bit slot duration T_b, as shown in Fig. 5.4(b) and (c). Also, for the convenience of discussion, we assume that the gain/phase saturate and recover with a constant slope, while with an exponential slope in reality. In view of the asymmetrically located SOA in the Sagnac loop, Fig. 5.4(b) and (c) also show that the counter-clockwise propagating signal is delayed by ΔT with respect to the clockwise propagating signal, when they enter the coupler C1.

---

[17] The subscript "s" of $G_s(t)$ in Eq.(2.30) is omitted.

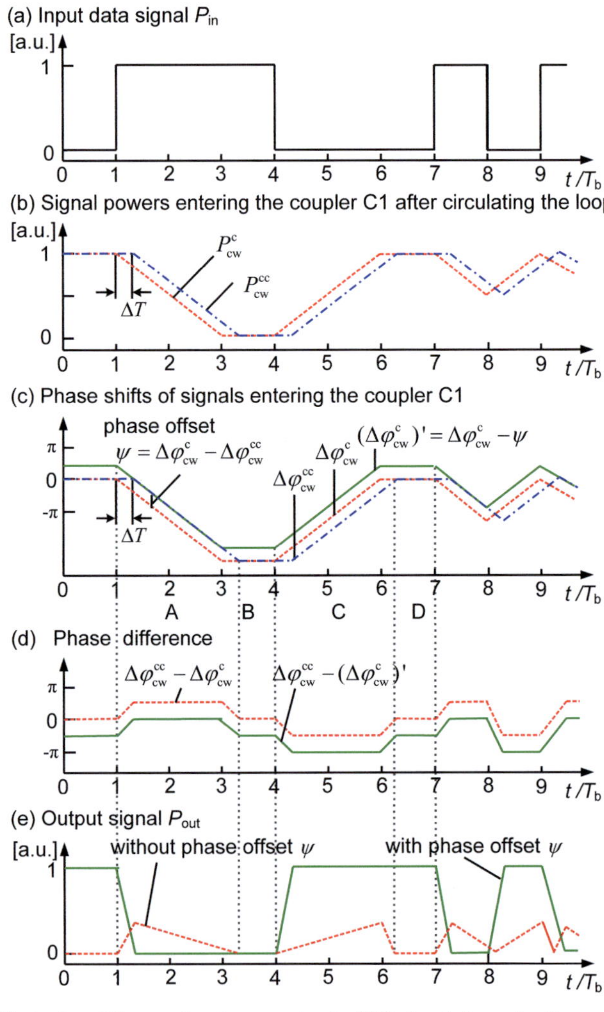

Fig. 5.4. Illustration of the wavelength conversion for NRZ signals through a Sagnac loop shown in Fig. 5.2. For an input signal (a), the P_{cw}^{c} and P_{cw}^{cc} and the phase shifts $\Delta\varphi_{cw}^{c}$ and $\Delta\varphi_{cw}^{cc}$ of two counterpropagating signals entering the coupler C1 are schematically shown as dotted and dashed lines in (b) and (c), respectively. In view of the asymmetrically located SOA in the Sagnac loop, the modulated counter-clockwise signal P_{cw}^{cc} is delayed by a time offset ΔT with respect to the modulated clockwise signal P_{cw}^{c}, when they entering the coupler C1. By detuning the polarization controller PC2, see Eq. (5.5), a proper phase offset ψ is added onto the clockwise propagating signal, solid line $(\Delta\varphi_{cw}^{c})'$ in (c). (d) shows phase difference between two counter-propagating signals without and with the phase offset ψ, dotted and solid lines. (e) shows output NRZ signals, where dotted and solid lines are results without and with the phase offset ψ. T_b is one bit slot duration.

In the following, we explain how both XGM and XPM are utilized to form a nice wavelength converted inverted NRZ signal. The inverted operation means that a bit 1 (mark) is converted to a bit 0 (space), or vice versa.

We first look at the phase difference inside the loop and the output signal out of the loop when there is not a phase offset ψ between two counter-propagating signals, dotted lines in Fig. 5.4(d) and (e). We have split the time interval between $t/T_b = 1$ to 7 into 4 time windows. In time windows A and C, where there are gain/phase saturation or recovery, the phase difference $\Delta\varphi_{cw}^{cc} - \Delta\varphi_{cw}^{c} \neq 0$ due to the time delay between two counter-propagating signals. As shown in from Eq. (5.5), this non-zero phase differences leads to constructive interferences and to non-zero output pulses, dotted line in Fig. 5.4(e). In time windows B and D, where gain/phase saturation or recovery finishes, the phase difference $\Delta\varphi_{cw}^{cc} - \Delta\varphi_{cw}^{c} = 0$. The output NRZ signal is then zero.

Now, we discuss the fact that with a phase offset ψ this wavelength conversion scheme can output an inverted NRZ signal. The phase difference and the output signal are depicted as solid lines in Fig. 5.4(d) and (e), respectively. An ideal phase offset ψ from PC2 should be set as follows. 1) On the one hand, the phase difference $\Delta\varphi_{cw}^{cc} - (\Delta\varphi_{cw}^{c})' = \Delta\varphi_{cw}^{cc} - \Delta\varphi_{cw}^{c} + \psi$ during the gain/phase saturation regime (e.g. time window A in Fig. 5.4(d)) should be compensated to be zero. 2) On the other hand, the phase difference $\Delta\varphi_{cw}^{cc} - (\Delta\varphi_{cw}^{c})'$ during the gain/phase recovery regime (e.g. time window C in Fig. 5.4(d)) is enlarged as well as the constructive interference is enhanced.

To output a nice inverted NRZ signal, the phase relation (i.e. XPM) should cooperate with the XGM effect. On the transitions depicted in Fig. 5.4(e), these two phase relations can perform a fast switching operation, while the gain saturations and the phase shifts depicted in Fig. 5.4(b) and (c) show much slower dynamics. In addition, as seen in time windows B and D in Fig. 5.4(d), the phase differences $\Delta\varphi_{cw}^{cc} - (\Delta\varphi_{cw}^{c})'$ are not zero any more. In fact, the phase difference is reset to the phase offset ψ. However, the output power in time window B is low due to the strong gain saturation, as shown in Fig. 5.4(b), while the output power in time window D is high since the gain is fully recovered. In total, the output signal is an inverted NRZ signal, Fig. 5.4(e), with respect to the input signal.

We should mention that a non-inverted NRZ signal can also be output from the Sagnac loop. It is achieved by detuning the phase offset such that the phase difference $\Delta\varphi_{cw}^{cc} - (\Delta\varphi_{cw}^{c})'$ is zero for the low level input pulses, e.g. time window C in Fig. 5.4(c). From Eq. (5.5), this phase detuning gives a destructive interference and in turn a low level output signal. On the other hand, the interference in the case of high level input pulse is now constructive and will give a relatively high level output signal. So a non-inverted signal is resulted. However, due to the gain saturation in the SOA for the high level input pulses, the extinction ratio of the non-inverted output signal is not as good as for the inverted signal.

The discussion above excludes the realistic carrier recovery dynamics, which plays an important role in the converting process in the SOA. In the realistic case, the ideal phase setting condition, where $\Delta\varphi_{cw}^{cc} - \Delta\varphi_{cw}^{c} + \psi$ is zero for input bit ones, is not hold either for multiple input pulses or even during a single bit period. In the following, to determine the requirements on an SOA to provide best results from a Sagnac loop, we have to include the carrier recovery dynamics.

5.2 Results with SOAs Having Various Carrier Recovery Times

The carrier recovery time of an SOA is related to the low-pass characteristics of the SOA. A larger carrier recovery time represents a smaller bandwidth of the SOA transfer function. A Sagnac interferometer is to differentiate the SOA's output and to restore the shape of the incident signal. At a required bit rate, a fixed Sagnac interferometer can only compensate the low-pass characteristics of SOAs with certain carrier recovery time. Intuitively, a Sagnac interferometer may not compensate the low-pass characteristics of a very slow SOA, and it may overcompensate the low-pass characteristics of a very fast SOA.

In this section, we will investigate the quality of the wavelength conversion using SOAs having various carrier recovery times in the Sagnac loop. The discussion is performed by using the SOA model given in Chapter 2. Thus, we take into account slow interband filling as well as fast intraband gain compression effects, e.g. spectral-hole burning (SHB), carrier heating (CH), and two-photon absorption (TPA), in the gain and phase dynamics.

5.2.1 Characteristics of Various SOAs

Applying the numerical SOA model from Chapter 2, we selected three SOAs with 10:90 carrier recovery times of ~ 76 ps, ~ 52ps and ~ 34 ps – but with identical gain spectra.

The carrier recovery time is derived from the pump-probe experiment schematically shown in Fig. 5.5(a). A cw-probe signal and a pump pulse at another wavelength are launched into the SOA. The recovery times cited above really are the phase recovery times of the SOAs and not the gain recovery times. The reason that we choose the phase recovery times for reference is because the gain dynamics usually comprises both the ultrafast dynamics associated with SHB and CH as well as the slow band-filling effect – whereas the phase recovery mostly is influenced by the band-filling dynamics. The slow band-filling effect is the relevant process in our case at 40 Gbit/s. Thus, the factor α_N associated with the band-filling effect in Eq. (2.37) is of importance.

The respective SOAs, SOA1, SOA2 and SOA3 with 10:90 band-filling recovery times of ~ 76 ps, ~ 52 ps and ~ 34 ps have been obtained by varying the differential gain and bias current of a generic SOA, Fig. 5.5(b). The cw-probe signal is at 1550 nm with a power level of 5 dBm and The pump pulse is at 1558 nm with a peak power of 14.77 dBm (30 maw) and a pulse width of 8.33 ps (full-width at half maximum). The length of the active region of the SOAs is 1 mm and other SOA parameters are based on experimental data. Details of the parameters are given in Appendix B – except for α_N that has been chosen to be a constant value 5.

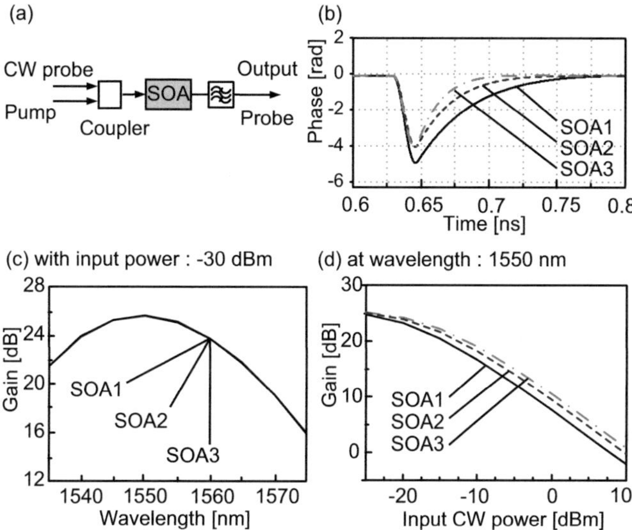

Fig. 5.5. (a) Schematic pump-probe experiment setup. (b) Phase recovery dynamics of three pre-pared SOAs. (c) Gain spectra and (d) gain saturation characteristics. The CW power in (c) is −30 dBm, while the wavelength of input CW signal in (d) is at 1550 nm.

While these three SOAs have different phase recovery times, their gain spectra are identical, Fig. 5.5(c). Further, associated with the recovery time, these three SOAs show different gain saturation characteristics, shown in Fig. 5.5(d). The input saturation power gradually increases from SOA1 to SOA3. This observation coincides with the definition of the saturation power in [2].

5.2.2 Wavelength Conversion at 40 Gbit/s

In this section, we will analyze the wavelength conversion with three respective SOAs at 40 Gbit/s. Their performances are analyzed by launching a noise loaded PRBS signal in the Sagnac loop and by evaluating the best Q-factors at the optimum operation point of each wavelength converter. We observed that the best 40 Gbit/s wavelength conversion is obtained with SOA2, having a recovery time of 52 ps.

Since the main interest in this work is the influence of the SOA dynamics, the power splitting ratios of the couplers used in the simulation are then fixed. Accordingly to those in an experiment [30], the power splitting ratio was chosen to be 50:50 for the coupler C1, while 80:20 for the coupler C2 (co-propagating signal transmission 80%, input data signal transmission 20%). The deviation of these ratios in the respective couplers may modify the requirements on the input powers but do not change the principle of the operation.

The input data signal used in the simulation is a noisy NRZ signal at an optical carrier of 1558 nm, which is modulated with a PRBS sequences of $2^8 - 1$ at 40 Gbit/s. The input data signal has an average power of 12 dBm and the CW signal has a power level of 3 dBm. The time delay ΔT has been set to 2 ps. All simulations have been carried out with the optical communication system simulator OptSim from RSoft, while the SOA model from Chapter 2 has been implemented.

Results with an SOA Having Moderate Recovery Speed

Now we give our simulation results with the SOA2, which has a moderate recovery speed shown in Fig. 5.5. An input signal launched into a Sagnac loop, Fig. 5.6(a), generates a transmitted and reflected signal at the respective outputs; see solid and dashed lines in Fig. 5.6(d). Note that only snapshots between 0.4 ns and 1.2 ns are shown. The transmitted signal is inverted with respect to the input signal and the transitions between marks and spaces are fast and follow the bit rate of the input signal.

The quality of the transmitted signal depends on the output powers and relative phase difference of the co- and counter-propagating signals, which we now look at more precisely. Note that the output signals in Fig. 5.6(b) are delayed with respect to the input signal in Fig. 5.6(a), because the signal transit time through the SOA is included in Fig. 5.6(b). Solid and dashed lines in Fig. 5.6(b) show the respective powers of the clockwise propagating signal P_{cw}^{c} and of the counter-clockwise propagating signal P_{cw}^{cc}, when they enter the coupler C1. One can see a strong pattern effect due to the slow recovery time. The nonlinear phase shift induced by the input signal onto the co- and counter-propagating signals are depicted on the upper half of Fig. 5.6(c). The amount of power that is mapped onto the output really depends on the phase difference $\Delta\varphi_{cw}^{cc} - (\Delta\varphi_{cw}^{c})'$, which is depicted on the right y-axis of Fig. 5.6(c). It can be noted that, at each time when the input signal transits from $0 \rightarrow 1$ (e.g. the time point (I) in Fig. 5.6(c)), the phase difference is becoming close to zero. On the other hand, at each time when the input signal transits from $1 \rightarrow 0$ (e.g. the time point (II) in Fig. 5.6(c)), the phase difference is larger. Whereas in all other cases, when there are a series of 0s or multiple 1s, the phase difference approaches the phase offset ψ set by PC2, shown as a dotted line in Fig. 5.6(c).

To set $\Delta\varphi_{cw}^{cc} - (\Delta\varphi_{cw}^{c})' \approx 0$ as the input signal transits from $0 \rightarrow 1$, e.g. at the time point (I) in Fig. 5.6, the phase offset ψ is chosen such that

$$\psi = -\min\left(\Delta\varphi_{cw}^{c} - \Delta\varphi_{cw}^{cc}\right), \forall P_{in} : 0 \rightarrow 1. \tag{5.8}$$

This phase offset from Eq. (5.8) ensures that the strongest gain depletion and in turn the largest phase shift induced by the input signal results in the lowest output power after the Sagnac loop. Except for this, the parameter γ in Eq. (5.6) is simplified to be 1 [55].

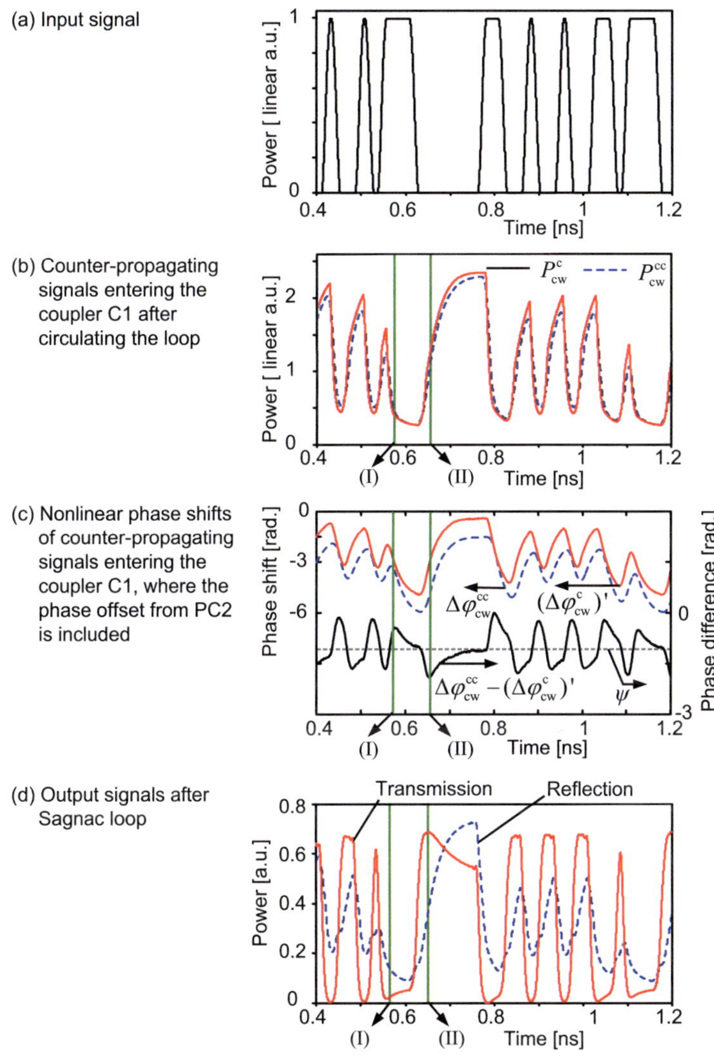

Fig. 5.6. Wavelength conversion from an input signal (a) to output signals (d) through a Sagnac loop with an SOA2, whose characteristics are referred to in Fig. 5.5. (b) shows the output power of the clockwise propagating signal P_{cw}^{c}, solid line, and of the counter-clockwise propagating signal P_{cw}^{cc}, dashed line, after the SOA2. Note that the output signals in (b) are delayed with respect to the input signal in (a), due to the transit time through the SOA. Left and right *y*-axes of (c) are the nonlinear phase shifts and the phase difference of two counter-propagating signals after the SOA2. The dotted line in (c) gives the phase offset ψ set by the PC2. All the signals are snapshots between 0.4 ns and 1.2 ns, where the solid lines and dashed lines in (d) are the transmitted and reflected signal powers respectively.

We now go into details of the fast switching processes. First we examine the situation when the output signal at the transmission port transits from a bit one to a bit zero, e.g. at the time point (I) in Fig. 5.6(d). At this time the phase difference $\Delta\varphi_{cw}^{cc} - (\Delta\varphi_{cw}^{c})' \approx 0$ ensures a destructive interference and switches off the device as from Eq. (5.5). Accompanying with the switch-off for the transmission port, the rest power of P_{cw}^{c} and P_{cw}^{cc} are guided to the reflection port, shown as the dashed line in Fig. 5.6(d). As subsequent input bits stay on the high level, e.g. the time slot between (I) and (II), the SOA gets more and more depleted and the device stays switched off. Meanwhile, the phase difference $\Delta\varphi_{cw}^{cc} - (\Delta\varphi_{cw}^{c})'$ gradually relaxes back to the initial phase offset ψ from Eq. (5.8), indicated with a dotted line in Fig. 5.6(c). However, even the device is gradually reset to the switch-on state; there is not much power at the output due to the strongly suppressed gain in the SOA. In short, the device now works mostly in the XGM mode.

Next, we examine the output signal at the transmission port transits from a bit zero to a bit one, e.g. at the time point (II) in Fig. 5.6. During this transition, the SOA recovers at a speed characterized by a carrier recovery time, which is usually longer than the duration of one bit. However, the phase difference $\Delta\varphi_{cw}^{cc} - (\Delta\varphi_{cw}^{c})'$ is almost -2 radians at this time point (II), seen in Fig. 5.6(c), and switches on the device. As a consequence, as shown in Fig. 5.6(d), most powers from P_{cw}^{c} and P_{cw}^{cc} are directed into the transmitted port (solid lines), thus speeding up the recovery. Further, as the SOA relaxes back to the initial carrier density level, the phase difference $\Delta\varphi_{cw}^{cc} - (\Delta\varphi_{cw}^{c})'$ also relaxes back to the initial phase offset ψ. Again, part of power from P_{cw}^{c} and P_{cw}^{cc} are directed into the reflection port. However, as seen in Fig. 5.6(d), not all the available power from P_{cw}^{c} and P_{cw}^{cc} are needed to give an ideal "1" state in the transmitted signal.

In summary, the Sagnac interferometer provides speed-up by means of XPM whenever there is a transition $0 \rightarrow 1$ or $1 \rightarrow 0$, but it operates as a pure XGM device in all other situations when there are long series of 0s or 1s in the output signal. In the stationary states after long series of 0s or 1s, we can see that the respective transmission $T_{0(1)}$ from Eq. (5.5) becomes

$$T_{0(1)} = \frac{G_{0(1)}}{2}\left[1 - \gamma\cos(\psi)\right], \tag{5.9}$$

where G_0 and G_1 are the stationary state gains after long series of 0s and 1s in the output signal, and $G_1 > G_0$. So, the extinction ratio between the bit 0 and the bit 1 of the output signal is ultimately determined by the XGM.

Comparison between SOAs Having Different Recovery Times

Both SOA XGM and XPM effects are related to the SOA carrier recovery dynamics. Indeed, a very fast SOA does not perform a good NRZ wavelength conversion through the Sagnac loop. The output signals from the Sagnac loop using a slower SOA1 and a faster SOA3 are shown in Fig. 5.7(a) and (b), respectively, where the solid lines and dashed lines are the transmitted and reflected signals. Except using different SOAs, same input signals are used and other parameters stayed unchanged.

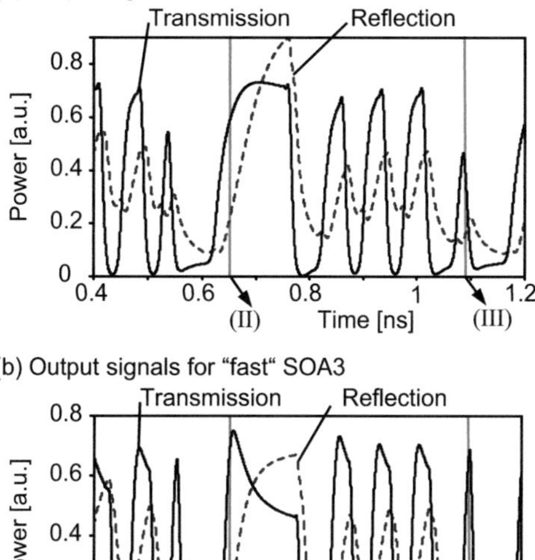

Fig. 5.7. Wavelength conversion through a Sagnac loop with (a) a "slow" SOA1 and (b) a "fast" SOA3, referred to in Fig. 5.5. All the signals are snapshots between 0.4 ns and 1.2 ns, where the solid lines and dashed lines in (a) and (b) are the transmitted and reflected output signals respectively.

1) When a slow SOA1 is used, the XPM effect is dominant. As the multiple input bit 0s comes in, the phase difference $\Delta\varphi_{cw}^{cc} - (\Delta\varphi_{cw}^{c})'$ relaxes slowly back to the initial phase offset ψ. This slow phase recovery gives a quite constant power level in the transmitted signal; see the solid line after the time point (II) in Fig. 5.7(a). However, if a single input bit 0 comes in, e.g. corresponding to an output bit 1 (solid line) at the time point (III) in Fig. 5.7(a), the pattern effect due to the slow carrier recovery dynamics is still the problem.

2) When a fast SOA3 is used, the XGM effect is dominant. Since the carrier density recovers back to its initial level quickly, the pattern effect does not trouble the single output bit 1, see solid line at the time point (III) in Fig. 5.7(b). However, as multiple input bit 0s set in, corresponding to the output 1s (solid line) after the time point (II) in Fig. 5.7(b), the phase difference $\Delta\varphi_{cw}^{cc} - (\Delta\varphi_{cw}^{c})'$ relaxes back much quickly (with respect to the case using SOA2 or SOA1) to the initial phase offset ψ. As a consequence, more than enough light is guided to the reflection port. Thus, in total, the output NRZ signal is not well formed.

Next, we want to find out the operation range of the input powers for three SOAs with various carrier recovery times. We vary the CW signal power P_{cw} from -3 to 9 dBm and the input data signal power P_{in} from 6 to 18 dBm. Along with the power variations, we set the time delay ΔT to be 2 ps. In fact, we found that this 2 ps time delay is an optimum.

Note that in reality the noise in the output wavelength converted signal originates mainly from SOA amplified spontaneous emission (ASE). In our simulation, instead of adding rate equations for ASE photon densities in the SOA model, an additive noise is included in the output signal via a linear booster amplifier behind the Sagnac loop. The noise figure[18] of this linear booster amplifier is set to be 5 dB. Other parameters stay unchanged.

As a performance judgement criterion, we use the Q-factor measurement. This measurement is an effective estimation of the *bit-error rate* (BER), which requires costly equipments in the practice and also needs much longer computational time in the simulation. We extract the Q-factor from eye diagram of the detected signal. We calculate the Q-factor improvement between the input data signal and the output wavelength converted signal, while the input data signal has a constant Q-factor of 15.9 dB. A positive Q-factor improvement (in dB) represents signal regeneration. Comparing this Q-factor improvement for different SOAs can lead to a fair judgement on their performances. The results are given as contour figures with respect to input powers in Fig. 5.8(a), (b) and (c) for the "moderate fast" SOA2, the slow SOA1 and the fast SOA3, respectively.

Comparing the results among for different SOAs in Fig. 5.8, we have following observations for

1) Different CW power. The SOA2 with a moderate recovery time has a best performance in the wavelength conversion, Fig. 5.8(a). It is also known that the effective carrier recovery time changes along with changing the CW power P_{cw}. So, for a very high or very low CW power, the SOA2 carrier recovery time is not proper for the NRZ wavelength conversion at 40 Gbit/s.

Increasing the CW power in the SOA1 or decreasing the CW power in the SOA3 can also make their effective carrier recovery time approaching that of SOA2. This could lead to a good performance. For the SOA3, as the CW power decreases to -3 dBm, the carrier recovery time increases to about 44 ps. The Q-factor improvement also reaches 1 dB, seen in Fig. 5.8(b). However, the ASE noise for low CW powers in the SOA3 should become higher in the reality and the noise figure will be above 5 dB as assumed for $P_{cw} = 3$ dBm. This 1 dB Q-factor improvement is then doubtful. For the SOA1, the available gain for the input signal is limited at high CW powers and the Q-factor improvement is also bad, seen in Fig. 5.8(c).

---

[18] All amplifiers degrade the signal-to-noise ratio (SNR) of the amplified signal because of spontaneous emission that adds noise to the signal during its amplification. The noise figure F_n of an amplifier is defined as the ratio between the SNRs of the input and output signals of the amplifier [3], $F_n = \dfrac{(\text{SNR})_{in}}{(\text{SNR})_{out}}$, where SNR refers to the electric power generated when the optical signal is converted into an electrical current. The output ASE power P_{ASE} in a bandwidth Δf is then $P_{ASE} = F_n(G-1)hf\Delta f$, where G is the amplifier gain of the booster amplifier [94].

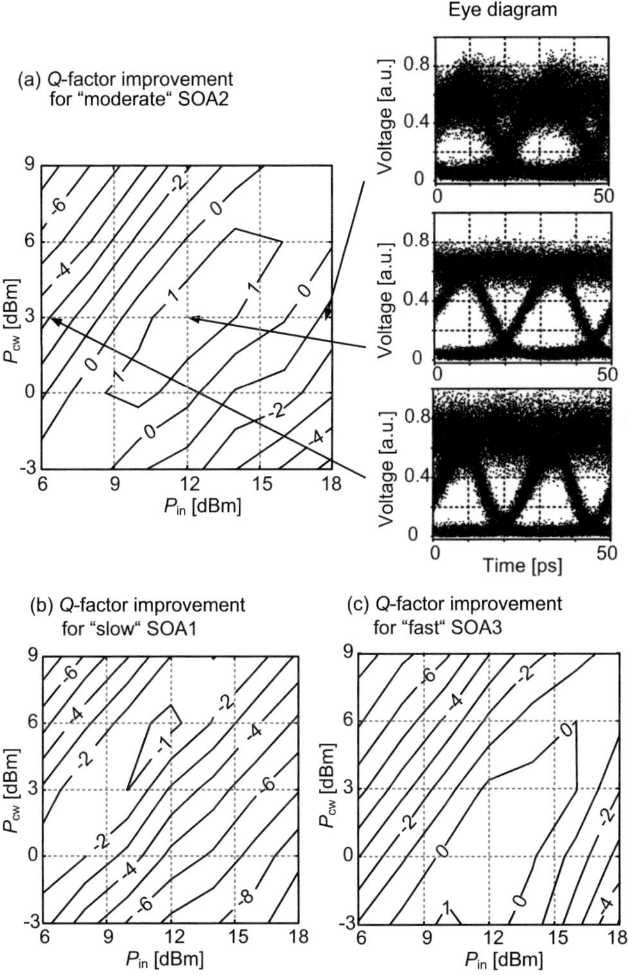

Fig. 5.8. *Q*-factor improvement of the **40 Gbit/s wavelength converted signals output from a Sagnac loop, using (a) a "moderate fast" SOA2, (b) a "slow" SOA1 and (c) a "fast" SOA3. The time offset ΔT in the Sagnac loop is 2 ps. The CW signal power P_{cw} varies from -3 to 9 dBm and the input data signal power P_{in} varies from 6 to 18 dBm. The *Q*-factor of the input data signal is 15.9 dB. In (a), the eye diagrams are shown for $P_{in} = 6$, 12 and 18 dBm, respectively, and $P_{cw} = 3$ dBm.**

2) Different input data power. Again, XGM and XPM induced by the input data signal P_{in} are of importance. On the one hand, a weak XGM and in turn a weak XPM at low input data power are not sufficient to perform the wavelength conversion. On the other hand, very high input data powers induce strong gain saturation in SOAs. The phase shift $\Delta\varphi_{cw}^{c}$ and $\Delta\varphi_{cw}^{cc}$ also

see this saturation effect and are equalized at very high input data power. In the consequence, the difference between $\Delta\varphi_{cw}^{c}$ and $\Delta\varphi_{cw}^{cc}$ disappears above certain input data powers. The largest phase difference $\Delta\varphi_{cw}^{cc} - \Delta\varphi_{cw}^{c}$, determining the phase offset ψ, happens as the input signal transits from a "0" to a "1" state. We found that, as the input signal power increases from 6 to 18 dBm while CW power is 3 dBm in Fig. 5.8(a), the phase difference $\Delta\varphi_{cw}^{cc} - \Delta\varphi_{cw}^{c}$ first increases and then decreases. The peak value of $\Delta\varphi_{cw}^{cc} - \Delta\varphi_{cw}^{c}$ happens at the input power of 12 dBm, which gives the best output signal as indicated in Fig. 5.8(a). In addition, if we take the carrier density dependent α_N-factor from Eq. (2.41) into consideration, the phase shift further decreases for very low carrier density.

Influence of the Linewidth Enhancement Factor

Now, we discuss the influence of the SOA linewidth enhancement α_N-factor (related the band-filling effect) on the wavelength conversion operation, since other linewidth enhancement factors accounting for SHB and CH play minor role in the phase dynamics. An ideal α_N-factor is such that the phase difference $\Delta\varphi_{cw}^{cc} - \Delta\varphi_{cw}^{c} + \psi$ is nearly 0 for a single output bit zero or multiple output bit zeros (destructive interference), and nearly π for a single output bit one or multiple output bit ones (constructive interference). However, the phase difference $\Delta\varphi_{cw}^{cc} - \Delta\varphi_{cw}^{c} + \psi$ for multiple bit zeros or ones is not a constant. In fact, due to the carrier recovery dynamics, the phase difference resets to the phase offset ψ, see Fig. 5.6(c). So, there is an optimum phase offset ψ, which has to balance the destructive and constructive interference for multiple bit zeros and ones.

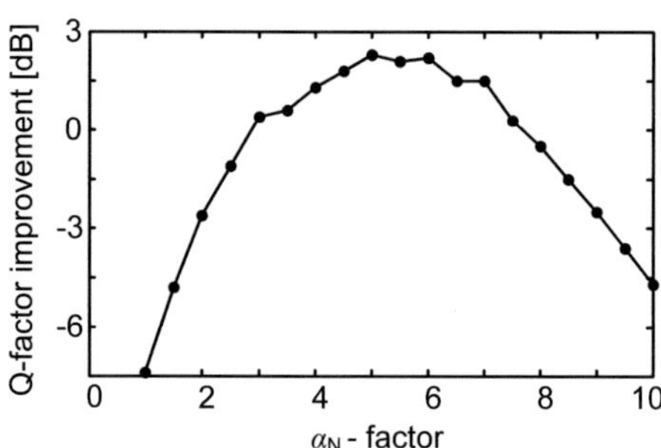

Fig. 5.9. *Q*-factor improvement of the output wavelength converted signals as a function of the SOA α_N-factor.

We run the simulations for different SOA α_N-factors. The time delay between the two counter-propagating waves is set to be 2.5 ps, while the CW signal power P_{cw} is 3 dBm and the input data signal power $P_{in} = 12$ dBm. We plot the *Q*-factor improvement of the output wavelength converted signal as a function the α_N-factor in Fig. 5.9. We see that at the chosen power levels the optimum α_N-factor is about 5.

5.2.3 Wavelength Conversion at 80 Gbit/s

Now we analyze the wavelength conversion at 80 Gbit/s with the SOA2 and SOA3 in the Sagnac loop, respectively. The CW signal power P_{cw} varies from -3 to $+9$ dBm and the input data signal power P_{in} varies from 6 to 18 dBm, whose Q-factors are fixed to about 14.8 dB. At 80 Gbit/s, the data pulse width (a FWHM of 4.17 ps) is narrower and the pulse slope is steeper than that at 40 Gbit/s. As a consequence, a smaller time offset ΔT is enough to give a required phase difference between two counter-propagating signals. In fact, in the simulation, we found that the optimum time offset ΔT decreases now to be 1 ps. Other parameters stay unchanged. The SOA2 and SOA3 have a recovery time (10% to 90%) of ~ 52 ps and ~ 34 ps, respectively, as shown in Fig. 5.5(b). Both recovery times are larger than the double of the bit duration (12.5 ps at 80 Gbit/s), so these two SOAs are not fast enough for the operation at 80 Gbit/s. However, we can still find how the SOA carrier lifetime influences the wavelength conversion result.

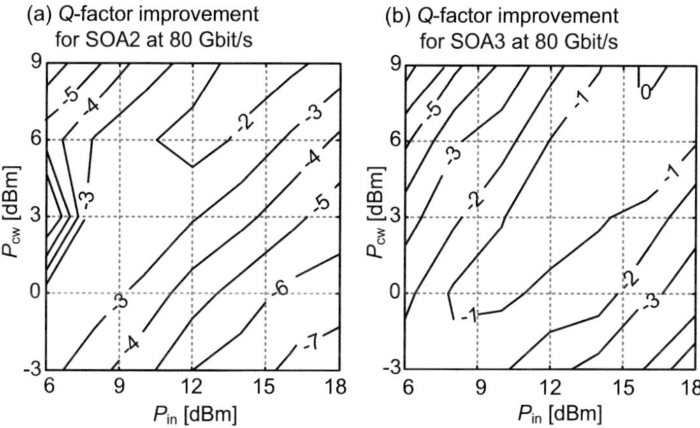

Fig. 5.10. *Q*-factor improvement of the output wavelength converted signals after a Sagnac loop at 80 Gbit/s, using (a) a "moderate fast" SOA2, (b) a "fast" SOA3. The time delay ΔT in the Sagnac loop is 1 ps. The CW signal power P_{cw} varies from $-$ 3 to 9 dBm and the input data signal power P_{in} varies from 6 to 18 dBm. The *Q*-factor of the input data signal is 14.8 dB.

The Q-factor improvements of the 80 Gbit/s wavelength converted signals are given in the contour figures in Fig. 5.10, where (a) is using a "moderate fast" SOA2 and (b) using a "fast" SOA3. From the results in Fig. 5.10, we hardly see any positive Q-factor improvements in a large power range. However, for a $P_{cw} = 9$ dB and $P_{in} = 16$ dB, a Q-factor improvement of 0 dB is observed by using SOA3. This is due to the fact that the recovery time of SOA3 is sped up to about 30 ps for $P_{cw} = 9$ dB. So, if an SOA with a recovery time of 25 ps is used, larger Q-factor improvement can be expected.

5.3 Results with Differently Long SOAs

As discussed in section 3.3.2, the gain saturation characteristics for two counter-propagating waves through a longer SOA are considerably different. Thus, a Sagnac interferometer may compensate the SOA low-pass characteristics for one output signal after the SOA, but not for the counter-propagating one. In this section, we will compare the wavelength conversion results with differently long SOAs used in the Sagnac loop.

5.3.1 Characteristics of Various SOAs

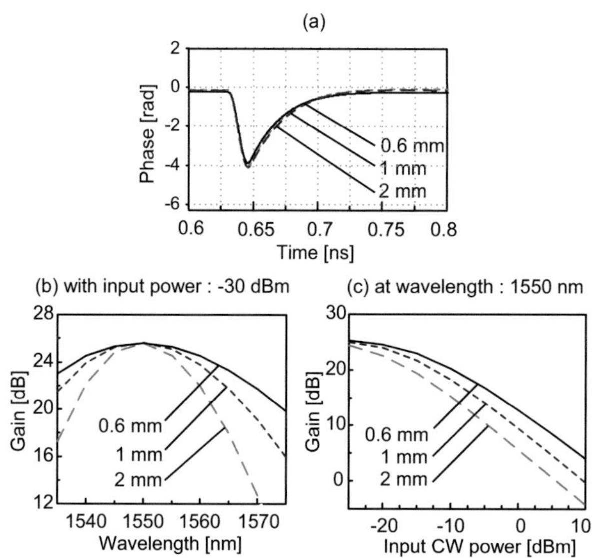

Fig. 5.11. Characteristic of SOAs with a length of 0.6 mm, 1 mm and 2 mm. (a) Phase recovery dynamics of prepared SOAs from simulation shown in Fig. 5.5(a). (b) Gain spectra and (c) gain saturation characteristic. The CW power in (b) is −30 dBm, while the wavelength of input CW signal in (c) is at 1550 nm.

In the numerical experiment shown in Fig. 5.5(a), we prepared two more SOAs having an active region length of 0.6 mm and 2 mm respectively. In the following we term the 0.6 mm long SOA as SOA4 and 2 mm long SOA as SOA5. Fig. 5.11 shows the characteristics of these two SOAs. They are compared with the characteristics of the SOA2 having an active region length of 1 mm. For same input powers used in section 1, the carrier recovery times of three SOAs are kept to be ~ 52 ps, Fig. 5.11(a). The same carrier recovery time is achieved by varying the bias current applied, according to SOA lengths. However, the gain spectra of these three SOAs are different. In fact, a long SOA can be understood as a cascading of several short SOAs. Thus, as shown in Fig. 5.11(b), the longer the SOA, the narrow the gain spectrum is. In addition, Fig. 5.11(c) compares the gain saturation characteristics of these three SOAs.

We can observe that for same saturation levels, a weaker input power is needed for a longer SOA, i.e. a longer nonlinear interaction length.

5.3.2 Wavelength Conversion at 40 Gbit/s

With SOA4 and SOA5 prepared above, we simulate the wavelength conversion at 40 Gbit/s. Results are shown in Fig. 5.12. The solid and dashed lines in Fig. 5.12(a) are the clockwise and counter-clockwise propagating signals P_{cw}^{c} and P_{cw}^{cc} after SOA4 or SOA5, respectively.

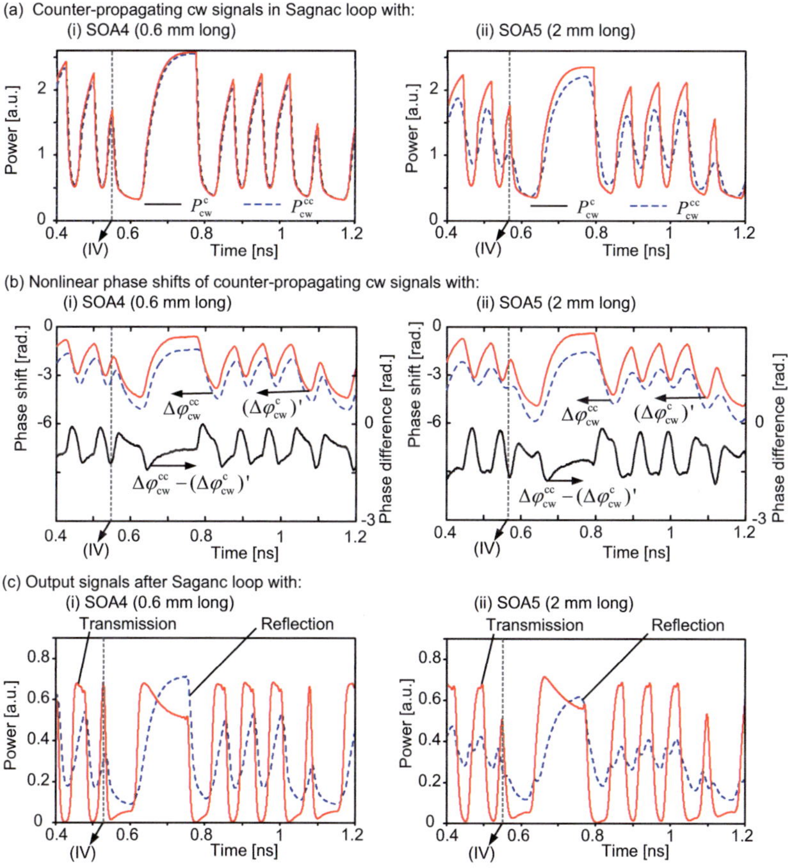

Fig. 5.12. (a) shows the output power of the clockwise and counter-clockwise propagating signals P_{cw}^{c}, solid line, and P_{cw}^{cc}, dashed line, after the SOA4 or the SOA5, which has an active length of 0.6mm and 2 mm respectively. (b) shows the nonlinear phase shift and the phase difference of two counter-propagating signals after the SOA. (c) Wavelength converted signals through the Sagnac loop. The solid and dashed lines in (c) are the transmitted and reflected signals respectively.

Fig. 5.12(b) shows the nonlinear phase shift $\left(\Delta\varphi_{cw}^{c}\right)'$ and $\Delta\varphi_{cw}^{cc}$ of two counter-propagating signals after the SOA, and the phase difference in-between (right y-axis). Fig. 5.12(c) shows wavelength converted signals through the Sagnac loop, where the solid and dashed lines are the transmitted and reflected signals at the output port of the coupler C1 shown in Fig. 5.2, respectively. All the signals are snapshots between 0.4 ns and 1.2 ns. The input data signal has an average power of 14 dBm for SOA4 and 10 dBm for SOA5. The CW power is 3 dBm. The time offset ΔT is 2 ps and the α_N-factor is 5. Other parameters are unchanged.

Comparing the output signals by using SOA2 (Fig. 5.6(d)), SOA4 and SOA5 (Fig. 5.12(c)), we find that using a longer SOA5 is not a good choice for the Sagnac loop based wavelength conversion. The reason lies in the asymmetric gain saturation and phase shift of two counter-propagating signals after the SOA. On one hand, the clockwise propagating signal P_{cw}^{c} co-propagates with the input data signal through the SOA and experiences the XGM and the XPM, which are proportional to the integral over the input data signal [19] and [31]. This proportionality is also given in Appendix E. On the other hand, the gain saturation and phase shift for the counter-clockwise propagating signal P_{cw}^{cc} are not only proportional to the integral over the input data signal, but also proportional to the integral over the SOA length [31]. Indeed, in a longer SOA, the gain saturation for P_{cw}^{cc} affects over a large time scale. For instance, dashed lines at the time point (IV) in right figure of Fig. 5.12(a) and (b), the power gain and the phase shift of P_{cw}^{cc} have not yet recovered, as the next pulse comes. As a consequence, the output power for such a bit (time point (IV)) is also low, see in right figure of Fig. 5.12(c). Note that the time points (IV) for SOA4 and SOA5 in Fig. 5.12(c) are at different times, due to different signal transit times through SOA4 and SOA5.

The advantage by using a shorter SOA is also verified by evaluating the Q-factor improvements of the wavelength converted signals, shown in Fig. 5.13(a) for SOA4 (0.6 mm long) and Fig. 5.13(b) for SOA5 (2 mm long). The input powers vary as given by the axes. In Fig. 5.13(a) for SOA4, there is a larger area of a Q-factor improvement above 1 dB, and even Q-factor improvements above 1.5 dB is also gotten. In comparison with the case of using SOA2 (1mm long, Fig. 5.8(a)), the performance by using the SOA4 is better. Also, as shown in Fig. 5.13(b), there is not any positive Q-factor improvement by using the longer SOA5.

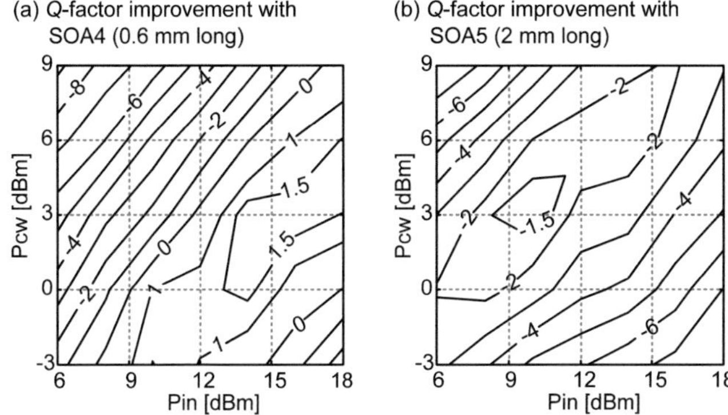

Fig. 5.13. *Q*-factor improvement of the wavelength converted signals after a Sagnac loop using (a) a shorter SOA4 (0.6 mm long), (b) a longer SOA5 (2 mm long). The time delay ΔT in the Sagnac loop is 2 ps. The CW signal power P_{cw} varies from -3 to 9 dBm and the input data signal power P_{in} varies from 6 to 18 dBm. The *Q*-factor of the input data signal is 15.9 dB.

In conclusion, we analyzed the wavelength conversion process in the SOA-based Sagnac loop. From more accurate simulations, we found that there are optimums for the CW power, the input data signal power and the SOA carrier recovery time. The XGM and XPM effects in the SOA should collaborate in a proper way for the NRZ wavelength conversion. The simulation result also reveals that an SOA with a carrier recovery time between 2 and 3 times of one bit duration gives a best output NRZ signal. Beside this carrier recovery time requirement, short SOAs are rather preferred in the Sagnac loop. This observation lies in the fact that material responses experienced by the signals from different directions become symmetric in short SOAs.

Even the operation speeds analyzed in this work are 40 Gbit/s and 80 Gbit/s, the criteria of choosing a proper SOA is applicable at higher speed NRZ wavelength conversion based on the Sagnac loop.

6 Wavelength Conversion for Differential Phase-Shift Keying Signal

Published by P. Vorreau, A. Marculescu, J. Wang, G. Boettger, B. Sartorius, C. Bornholdt, J. Slovak, M. Schlak, C. Schmidt, S. Tsadka, W. Freude, J. Leuthold, "Cascadability and regenerative properties of SOA all-optical DPSK wavelength converters," *IEEE Photon. Technol. Lett.*, vol. 18, no. 18, pp. 1970 - 1972, Sep. 2006

Differential phase-shift keying (DPSK) seems to be the most promising phase-modulation format in point-to-point and next generation meshed transparent networks, because of its robustness towards nonlinear impairments [18]. Especially, when a balanced detection is used, DPSK signal requires about 3 dB lower optical signal-to-noise ratio (OSNR) than an on-off-keying (OOK) to reach a given bit-error probability, also called as bit-error rate (BER). This advantage directly turns out to be a doubled transmission distance.

In view of the significance of the DPSK format for next generation systems, it would be highly desirable to have small-footprint, low power-consuming all-optical DPSK regenerators and wavelength converters. Indeed, solutions exploiting fiber-nonlinear effects [17] or the four-wave mixing (FWM) effect in semiconductor optical amplifiers (SOAs) [48] have been tested. Unfortunately, all of these solutions only offer a restricted wavelength conversion range and require high input power levels. Recently, a Sagnac [21] and a Mach-Zehnder interferometer (MZI) device [33] exploiting the more efficient SOA-based cross-phase modulation effect have been demonstrated and their regenerative potential has been shown. Yet, none of these solutions also showed retiming regeneration, neither the cascadability. This is of particular importance, since the DPSK format is most likely to be used in systems at bit rate of 40 Gbit/s and above, where retiming is an issue.

In this Chapter, we will introduce an SOA based MZI all-optical DPSK wavelength converter and discuss how these devices can be cascaded an arbitrary number of times. We also demonstrate in an experiment the feasibility and limits of this type of DPSK wavelength converter.

6.1 Configuration and Operation Principle

The DPSK wavelength conversion system is shown in Fig. 6.1. This setup comprises a DPSK transmitter, followed by the wavelength converter and the balanced receiver with differential decoding. The wavelength converter consists of a delay interferometer (DI) stage and a MZI stage with SOAs in the two arms, and an optical source with a target wavelength λ_{cnv} to which the DPSK signal at λ_{in} will be converted.

The operation principle of the all-optical wavelength converter is visualized in Fig. 6.1. An incoming DPSK formatted signal E_{in} at λ_{in} is transformed by the DI stage into an on-off keying (OOK) data E_{up} and an inverted OOK data E_{low}. These on-off signals are then used for controlling the SOAs. They are injected with exact time correlation into the two control inputs of a Mach Zehnder interferometer (MZI) with integrated SOAs [76]. A converting (i.e. probe) signal E_{cnv} at a wavelength λ_{cnv} (a clock signal or continuous wave (cw)) is introduced into the MZI and split up into the two arms of the MZI. The two signals on the respective arms are $E_{cnv,up}$ and $jE_{cnv,low}$. Note that a phase factor j will be added to the signal that couples into the cross output of the coupler. The relative phase and amplitude between $E_{cnv,up}$ and $jE_{cnv,low}$ is then controlled by the mark and space bit of the OOK and the inverse OOK via the SOA, very similar to the push-pull operation of electrically controlled MZI modulators used in DPSK transmitters [17].

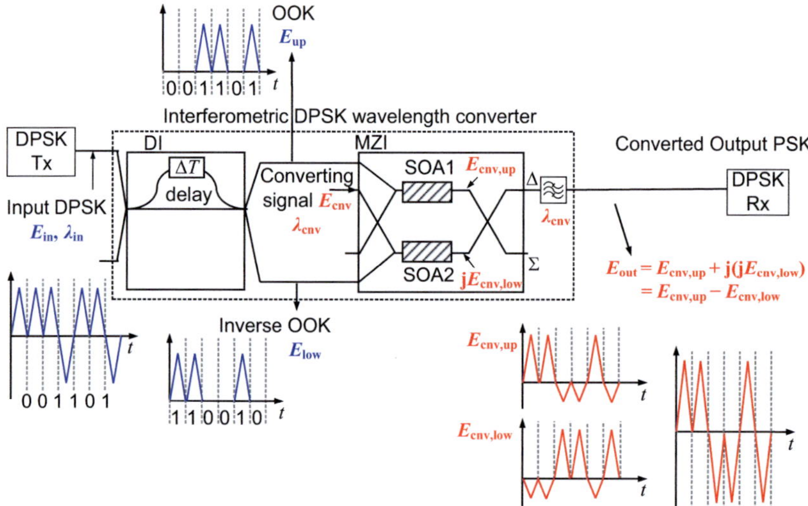

Fig. 6.1. Scheme of DPSK wavelength conversion with SOAs in the Mach-Zehnder interferometer (MZI) arms. The DPSK signal E_{in} is demodulated in the delay interferometer (DI, time delay equals the bit period T_b) resulting in an OOK and an inverse OOK signal. A phase factor j will be added to a signal that couples into the cross output of the 4-port couplers of the MZIs. A signal E_{cnv} at a new "converted" wavelength λ_{cnv} passes both SOAs resulting in the fields $E_{cnv,up}$ and $jE_{cnv,low}$ in the upper and lower MZI arms. At the difference output port Δ a filter selects the converted wavelength λ_{cnv} resulting in a signal E_{out}, which is sketched along with $E_{cnv,up}$ and $E_{cnv,low}$.

Fig. 6.1 shows schematically how the two amplitudes $E_{cnv,up}$ and $E_{cnv,low}$ in the upper and lower arm of the MZI change when an input OOK signal is introduced into the upper or lower arm of the MZI. For instance, if the input OOK signal is strong enough to induce a π phase shift, the amplitude of the converted signal in the respective arm is not only suppressed due to XGM, but also its phase is flipped due to XPM. The modulated $E_{cnv,up}$ and $jE_{cnv,low}$ are recombined at the difference output port Δ of the MZI. The output signal of the MZI passes a filter centered at λ_{cnv}. This results in a wavelength-converted output signal E_{out} at λ_{cnv}. For the logical levels "1" and "0" in the input DPSK signal, the converted signals have identical magnitudes, but opposite signs. In fact, the converted signal is a PSK-encoded signal. This means that the all-optical wavelength converter changes the coding scheme. In order to undo this new encoding, not only a balanced receiver at the output but a differential receiver with a differential encoder should be used. This is usually in the electrical domain.

6.2 Experiment Setup and Results near 40 Gbit/s

The wavelength conversion experimental setup and results are depicted in Fig. 6.2. The setup comprises a push-pull DPSK transmitter, followed by the all-optical wavelength converter and a balanced receiver with a differential encoding stage.

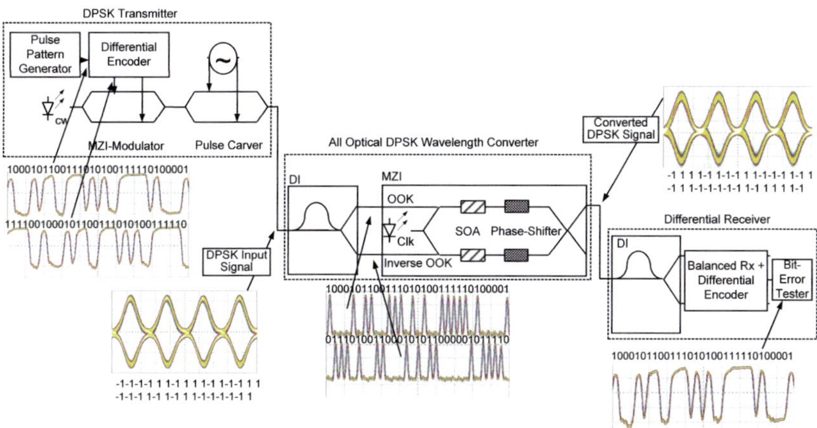

Fig. 6.2. Setup with DPSK transmitter, all-optical wavelength converter and balanced receiver with encoder, [81]. The two eye diagrams show the DPSK signal entering and leaving the all-optical wavelength converter. For the demonstration we reduced the bit rate to 31 Gbit/s because of pattern dependence at 40 Gbit/s.

For the experiment we used a 3 dBm input signal at $\lambda = 1545$ nm. The clock signal was set to $\lambda = 1563$ nm with a power level of 12.5 dBm. All powers are measured in the fiber in front of the device. A PRBS sequence of $2^7 - 1$ was chosen.

A 31 bit sequence and the eye diagrams at the various stages of the setup are shown at the bottom of Fig. 6.2. The eye diagrams of the DPSK signal before and behind the all-optical wavelength converter are clear and open. The converted signal shows some pattern dependent noise which is due to the limited recovery time of the SOAs. This pattern dependence limited operation to 40 Gbit/s. While error free operation at 40 Gbit/s was possible, the pattern dependence decreased when reducing the bit rate to 31 Gbit/s and it completely disappeared for a bit rate of 25 Gbit/s. The bit-error rate versus received signal is plotted in Fig. 6.3. A small penalty of 1 dB for DPSK wavelength conversion is found at 31 Gbit/s. No error floor was observed.

The eye pattern and the bit sequences at the bottom of Fig. 6.2 show how the input pattern undergoes various transitions. The NRZ input-pattern is plotted on the left side. It is then differentially encoded in order to produce the 33%-RZ-DPSK signal at the output of the transmitter stage.

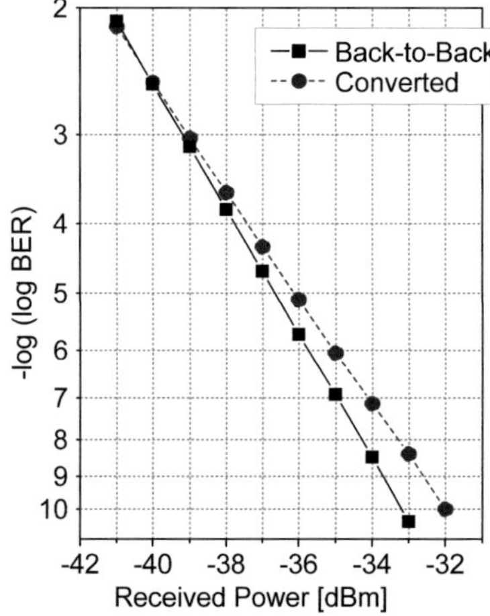

Fig. 6.3. Bit-error rate of the wavelength converted DPSK signal at 31 Gbit/s.

In the wavelength converter, the signal is first guided into the DI, which produces an OOK and inverse OOK pattern at the outputs. The new phase-encoded and wavelength converted signal is then formed in the MZI-stage. The phase and amplitude information – as annotated at the bottom of the eye diagrams before and after the converter in Fig. 6.2 – shows that the all-optical wavelength converter changes the coding scheme. In order to undo this new encoding we not only use a simple balanced receiver at the output but a differential receiver with a differential encoder. This provides exactly the same output pattern as fed into the system (see bit patterns in Fig. 6.2).

6.3 Cascadability

Due to the various bit-pattern transformations at different stages shown at the bottom of Fig. 6.2, it is not obvious if cascading the all-optical DPSK wavelength converter signal is possible at all. Here we show that an arbitrary number of all-optical wavelength converters can be cascaded. What matters is that there are an identical number of delay interferometer stages and differential encoders in the transmission link. The two elements, however, can be arranged in any order.

To prove this statement we first note, that the DI performs the inverse operation of the differential encoder. The logic operation performed by the two devices is depicted in Fig. 6.4.

(a) Differential Encoder

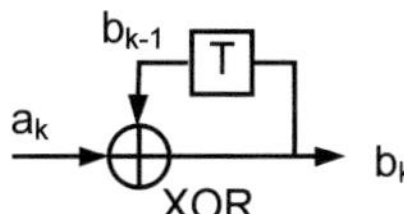

(b) Delay-Interferometer (DI)

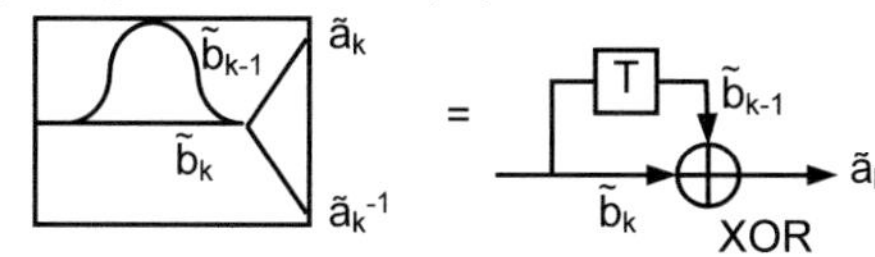

Fig. 6.4. (a) Logic operation performed by the differential encoder (typically an electronic device built into the receiver or transmitter). (b) Operation performed by optical delay interferometer (DI) and its logic analogue.

Mathematically, the differential encoder and the delay interferometer perform the following operations

$$b_k = b_{k-1} + a_k \;\; ; \tilde{a}_k = \tilde{b}_k + \tilde{b}_{k-1},$$ (6.1)

with the "+" obeying a modulo 2 algebra, i.e.

$$0+1 = 1+0 = 1 \text{ and } 0+0 = 1+1 = 0,$$ (6.2)

and a_k, b_k, $\tilde{a}_k$, $\tilde{b}_k \in \{0,1\}$. One now can show by substitution, that the output of a differential encoder followed by a DI returns identity if b_k is identical to $\tilde{b}_k$,

$$\tilde{a}_k = \tilde{b}_k + \tilde{b}_{k-1} = b_{k-1} + a_k + \tilde{b}_{k-1} = a_k.$$ (6.3)

The same is true for the reverse. Now, as in a transmission link with an arbitrary order of differential encoders and DIs one can always form consecutive pairs, that cancel each others operation. It thus follows that one always gets identity as long as there are the same number of differential encoders and DIs in the link.

112

7 Summary and Outlook

Summary

This work investigates the pattern effect mitigation techniques used in wavelength conversion with semiconductor optical amplifiers (SOAs). To understand the SOA behavior in wavelength conversion, an SOA model was established based on a rather simple set of rate equations for inter- and intraband carrier dynamics. The intraband carrier dynamics modeled in this work are carrier heating and spectral hole burning. The rate equations are simplified by assuming charge neutrality and neglecting the longitudinal and transverse carrier diffusion. Also, the group velocity dispersion was not taken into account. The traveling wave equations were used to describe the bi-directional pulse propagations through the SOA.

In most publications, while the gain dynamics are modeled via the rate equations, the phase dynamics in an SOA are modeled with the help of a constant linewidth enhancement factor, also called α-factor. However, the constant α-factor is valid only over a limited spectral range and for sufficiently small carrier density modulations in the SOA. In a wavelength conversion experiment, we observed that the cross-gain modulation (XGM) showed a different evolution to that of cross-phase modulation (XPM). First of all, the phase evolution in the experiment showed hardly any fast recovery effects, while the phase evolution derived from a constant α-factor did show a fast recovery. Therefore, the α-factor during the XGM experiment is not constant. Second, we also observed that the XPM lags behind the XGM by ~2 picoseconds for an input pulse width of 3 picoseconds. So, to properly model the phase dynamics in wavelength conversion, we used α-factors for individual carrier dynamics. Especially, a novel parameterization for the α-factor related to interband carrier dynamics were introduced. Inclusion of this model leads to excellent simulation results in agreement with 160 Gbit/s experiment results.

To relate the total phase change with the measured gain change of the output signal after an SOA, we introduced an effective α-factor. Its dynamics in wavelength conversion experiments were investigated. In the experiment and simulation, this effective α-factor varies strongly with time, different to a constant α-factor assumed in [27]. It even takes on negative values for short periods of time. The period of a negative effective α-factor is related to the lag of XPM behind XGM. This delay has important consequences for ultrafast operation of all-optical devices based on SOAs. It actually means that not each scheme will be suited for ultrafast operation. However, the lag of the XPM can also be made use of as a pattern effect mitigation technique in wavelength conversion.

Firstly, a delay interferometer (DI) filter is used to mitigate the SOA-related pattern effect. The DI filter performs a differential operation to overcome the speed limitation due to slow carrier recovery in the SOA. In this work, a return-to-zero signal to non-return-to-zero (NRZ) signal wavelength conversion at 160 Gbit/s was achieved by using an SOA followed by a DI filter. Experiment results of this SOA-DI scheme fit to the simulation results very well. The choice of the DI filter presents a passband to the blue-shifted spectral component, while partially suppressing the red-shifted spectral component. As a result, both the signal in the gain saturated regime and the red-chirped signal in the gain recovery regime experience a low transmission through the SOA-DI. Thus, an inverted NRZ signal can be gotten.

Extending the optical filtering technique, we investigated the pattern effect mitigation techniques in wavelength conversion with an SOA followed by a pulse reformatting optical filter (PROF). The PROF scheme exploits the fast chirp effects in the converted signal after the SOA and uses both the red- and blue-shifted spectral components. With a careful design of the PROF, we demonstrated a non-inverted wavelength conversion in an experiment at 40 Gbit/s, while the pattern effect due to the slow carrier recovery in the SOA was successfully mitigated. The pattern effect mitigation technique demonstrated here is based on the fact that the red chirp and the blue chirp in the inverted signal behind an SOA have complementary pattern effects. If the two spectral components are superimposed by means of the PROF, then pattern effects can be successfully suppressed.

The SOA-related pattern effect can also be mitigated by using the Sagnac loop, whose role is to differentiate the SOA's output and to restore the shape of the incident signal. In this work, NRZ to NRZ wavelength conversion in an SOA-based Sagnac loop was also investigated. We discussed how operation speed and SOA properties, especially the SOA carrier recovery time, are related. We found that SOA recovery times can be neither too fast nor too slow for a particular operation speed. An optimal carrier recovery time is between two and three times of one bit slot duration. We also found that shorter SOAs are preferred in this wavelength conversion scheme.

As differential phase-shift keying (DPSK) signal seems to be the most promising phase-modulation format in next generation meshed transparent networks, all-optical DPSK wavelength converters are of high interest. In this work, we proposed a DPSK wavelength conversion scheme comprising a DI and an SOA based Mach-Zehnder interferometer (MZI). A regenerative wavelength conversion experiment for DPSK signal was demonstrated in the 40 Gbit/s range. A small penalty of 1 dB for DPSK wavelength conversion was found at 31 Gbit/s. Its cascadability was also discussed.

Outlook

The work described in this thesis can be extended in the following aspects:

1) The numerical model can be improved. As SOA-based all-optical wavelength conversions will be used at highest speed, e.g. above 320 Gbit/s, the pulse width of the optical signal is comparable short with respect to ultrafast nonlinear effects. For this instance, group velocity dispersion may need to be taken into account. Also, polarization dependent effects can be taken into account, possibly by introducing the polarization dependent confinement factor. As amplified spontaneous emission (ASE) is unavoidable in the SOA, to obtain more accurate quantitative agreements with the experimen-

tal results, the modeling for the ASE noise should be also improved in the current model.

2) In principle, the PROF scheme discussed in Chapter 4 can work at bit rates higher than 40 Gbit/s. In fact, the PROF scheme demonstrated wavelength conversion at 40 Gbit/s, with record low input signal powers. So, successful experimental demonstrations at high bit rates are very probable and well worth a try.

3) While the work presented in this thesis is based on cross-gain modulation and cross-phase modulation, four-wave mixing (FWM) effect in the SOA can also be utilized to perform wavelength conversion. Accordingly, the SOA model needs to be extended to investigate the FWM effect.

Reference

[1]   G. P. Agrawal and N. Anders Olsson, "Self-phase modulation and spectral broadening of optical pulses in semiconductor laser amplifiers," *IEEE J. Quantum Electron.*, vol. 25, No. 11, p2297-2306, Nov. 1989.

[2]   G. P. Agrawal and N. K. Dutta, *Semiconductor lasers*, New York: Van Nostrand Reinhold, 2nd Ed., 1993.

[3]   G. P. Agrawal, *Fiber-Optic Communications Systems*, New York: Wiley, 3rd Ed., 2002.

[4]   M. Asada and Y. Suematsu, "Density-matrix theory of semiconductor lasers with relaxation broadening model – Gain and gain-suppression in semiconductor lasers," *IEEE J. Quantum Electron.*, vol. QE-21, pp. 434-442, May 1985.

[5]   S. Bischoff, M. L. Nielsen, and J. Mørk, "Improving the all-optical response of SOAs using a modulated holding signal," *IEEE J. Lightw. Technol.*, vol. 22, pp. 1303-1308, May 2004.

[6]   Ö. Boyraz, P. Koonath, V. Raghunathan, and B. Jalali, "All-optical switching and continuum generation in silicon waveguides," *Opt. Express*, vol. 12, pp. 4094-4102, Aug. 2004.

[7]   N. Chi, J. Zhang, P. V. Holm-Nielsen, C. Peucheret, and P. Jeppesen, "Transmission and transparent wavelength conversion of an optical labeled signal using ASK/DPSK orthogonal modulation," *IEEE Photon. Technol. Lett.*, vol. 15, pp. 760–762, May 2003.

[8]   W. W. Chow, S. W. Koch, and M. Sargent III, *Semiconductor-Laser Physics*, Berlin, Germany: Springer-Verlag, 1994.

[9]   M. J. Connelly, *Semiconductor optical amplifiers*, Kluwer Academic Publishers, 2002.

[10]   S. Diez, C. Schmidt, R. Ludwig, H. G. Weber, K. Obermann, S. Kindt, I. Koltchanov, and K. Petermann, "Four-wave mixing in semiconductor optical amplifiers for frequency conversion and fast optical switching," *IEEE J. Select. Topics Quantum Electron.*, vol. 3, pp. 1131-1145, Oct. 1997.

[11]   C. Dorrer and I. Kang, "Simultaneous temporal characterization of telecommunication optical pulses and modulators by use of spectrograms," *Opt. Lett.*, vol. 27, pp. 1315-1317, Aug. 2002.

[12]   M. Eiselt, W. Pieper, and H. G. Weber, "SLALOM: Semiconductor laser amplifier in a loop mirror," *IEEE J. Lightw. Technol.*, vol. 13, pp. 2099-2112, Oct. 1995.

[13]   W. Freude, *Seminar: Allgemeine Differentialgleichung erster Ordnung*, University of Karlsruhe, June, 2001.

[14]   R. Giller, R. J. Manning, and D. Cotter, "Gain and phase recovery of optically excited semiconductor optical amplifiers," *IEEE Photon. Technol. Lett.*, vol. 18, pp. 1061-1063, May 2006.

[15] F. Ginovart and J. C. Simon, "Semiconductor optical amplifier length effects on gain dynamics," *J. Phys. D, Appl. Phys.*, vol. 36, no. 13, pp. 1473-1476, 2003.

[16] F. Girardin, G. Guekos, and A. Houbavlis, "Gain recovery of bulk semiconductor optical amplifiers," *IEEE Photon. Technol. Lett.*, vol. 10, pp. 784-786, Jun. 1998.

[17] A. Gnauck, G. Raybon, S. Chandrasekhar, J. Leuthold, C. Doerr, L. Stulz, A. Agarwal, S. Banerjee, D. Grosz, S. Hunsche, A. Kung, A. Marhelyuk, D. Maywar, M. Movassghi, X. Liu, C. Xu, X. Wei, and D. M. Gill, "2.5 Tb/s (64×42.7 Gb/s) transmission over 40×100 km NZDSF using RZ-DPSK format and all-Raman-amplified spans", in *Techn. Digest of Optical Fiber Communication Conf. (OFC) 2002*, FC2.1-FC2.3, Anaheim, USA, 2002.

[18] A. H. Gnauck and P. Winzer, "Optical phase-shift-keyed transmission," *IEEE J. Lightw. Technol.*, vol. 23, pp. 115-130, Jan. 2005.

[19] E. Granot, R. Zaibel, N. Narkiss, S. Ben-Ezra, H. Chayet, N. Shahar, S. Sternklar, S. Tsadka, and P. R. Prucnal, "Tunable all-optical signal regenerator with a semiconductor optical amplifier and a Sagnac loop: principles of operation," *J. Opt. Soc. Am. B*, vol. 22, pp. 2534-2541, Dec. 2005.

[20] G. Grau and W. Freude, *Optische Nachrichtentechnik*, 3. Ed. Berlin: Springer Verlag 1991, University reprint, 2005.

[21] V. S. Grigoryan, M. Shin, P. Devgan, J. Lasri and P. Kumar, "SOA-based regenerative amplification of phase-noise-degraded DPSK signals: dynamic analysis and demonstration," *IEEE J. Lightw. Technol.*, vol. 24, pp. 135-142, Jan. 2006.

[22] Y. Guo, C. K. Kao, E. H. Li, and K. S. Chiang, *Nonlinear Photonics: Nonlinearities in Optics, Optoelectronics and Fiber Communications*, Chinese University Press and Springer-Verlag Berlin, 2002.

[23] K. L. Hall, J. Mark, E. P. Ippen, and G. Eisenstein, "Femtosecond gain dynamics in InGaAsP optical amplifiers," *Appl. Phys. Lett.*, vol. 56, pp. 1740-1742, Apr. 1990.

[24] K. L. Hall, G. Lenz, E. P. Ippen, U. Koren, and G. Raybon, "Carrier heating and spectral hole burning in strained-layer quantum-well laser amplifiers at 1.5 μm," *Appl. Phys. Lett.*, vol. 61, pp. 2512-2514, Nov. 1992.

[25] K. L. Hall and K. A. Rauschenbach, "100-Gb/s bitwise logic," *Opt. Lett.*, vol. 23, pp. 1271–1273, Aug. 1998.

[26] K. L. Hall, E. R. Thoen, and E. P. Ippen, *Nonlinear Optics in Semiconductors II, Semiconductors and Semimetals*, San Diego, CA: Academic, 1998, vol. 59.

[27] C. H. Henry, "Theory of the linewidth of semiconductor lasers," *IEEE, J. Quantum Electron.*, vol. QE-18. pp. 259-264., Feb. 1982.

[28] M. Y. Hong, Y. H. Chang, A. Dienes, J. P. Heritage, P. J. Delfyett, S. Dijaili, and F. G. Patterson, "Femtosecond self- and cross-phase modulation in semiconductor laser amplifiers," *IEEE J. Select. Topics Quantum Electron.*, vol. 2, pp. 522-539, Sept. 1996.

[29] D. C. Hutchings and E. W. Van Stryland, "Nondegenerate two-photon absorption in zinc blende semiconductors," *J. Opt. Soc. Am. B*, vol. 9, pp. 2065-2074, Nov. 1992.

[30] S. L. Jansen, H. Chayet, E. Granot, S. Ben Ezra, D. den van Borne, P. M. Krummrich, D. Chen, G. D. Khoe and H. De Waardt, "Wavelength conversion of a 40-Gb/s NRZ signal across the entire C-band by an asymmetric Sagnac loop," *IEEE Photon Technol. Lett.*, vol. 17, pp. 2137-2139, Oct. 2005.

[31] K. I. Kang, T. G. Chang, I. Glesk, and P. R. Prucnal, "Comparison of Sagnac and Mach-Zehnder ultrafast all-optical interferometric switches based on a semiconductor resonant optical nonlinearity," *Appl. Opt.*, vol. 35, pp. 417-426, Jan. 1996.

[32] I. Kang and C. Dorrer, "Measurements of gain and phase dynamics of a semiconductor optical amplifier using spectrograms," in OFC, Los Angeles, CA, 2004, Paper MF43.

[33] I. Kang, C. Dorrer, L. Zhang, M. Rasras, L. Buhl, A. Bhardwaj, S. Cabot, M.Dinu, X. Liu, M. Cappuzzo, L. Gomez, A. Wong-Foy, Y.F. Chen, S. Patel, D. T. Neilson, J. Jaques, and C. R. Giles, "Regenerative all-optical wavelength conversion of 40 Gb/s DPSK signals using a SOA MZI", in *Proc. 31st European. Conf. Optical Communications (ECOC)*, Th. 4.3.3, Glasgow, UK, 2005.

[34] C. Koos, L. Jacome, C. Poulton, J. Leuthold, and W. Freude, "Nonlinear silicon-on-insulator waveguides for all-optical signal processing," *Opt. Express*, vol. 15, pp. 5976-5990, May 2007.

[35] M. Kot and K. Zdansky, "Measurement of radiative and nonradiative recombination rate in InGaAsP-InP LED's", *IEEE J. Quantum Electron.*, vol. 28, pp. 1746-1750, Aug. 1992.

[36] H. A. Kramers, "Diffusion of Light by Atoms," *Atti. Congr. Internat. Fisici*, vol. 2, pp. 545-57, 1927.

[37] R. d. L. Kronig, "On the theory of dispersion of X-rays," *J. Opt. Soc. Am.*, vol. 12, pp. 547-557, June 1926.

[38] J. Leuthold, P. A. Besse, J. Eckner, E. Gamper, M. Dülk, and H. Melchior, "All-optical space switches with gain and principally ideal extinction ratios," *IEEE J. Quantum Electron.*, vol. 34, pp. 622-633, Apr. 1998.

[39] J. Leuthold, "Advanced Indium-Phosphide Waveguide Mach-Zehnder Interferometer All-Optical Switches and Wavelength Converters," *ser. Quantum Electronics. Kostanz, Germany: Hartung-Gorre*, vol. 12, 1999.

[40] J. Leuthold, M. Mayer, J. Eckner, G. Guekos, H. Melchior and Ch. Zellweger, "Material gain of bulk 1.55 μm InGaAsP/InP semiconductor optical amplifiers approximated by a polynomial model," *J. Applied Physics*, vol. 87, no. 1, 2000.

[41] J. Leuthold, C. H. Joyner, B. Mikkelsen, G. Raybon, J. L. Pleumeekers, B. I. Miller, K. Dreyer, and C. A. Burrus, "100-Gb/s all-optical wavelength conversion with integrated

SOA delayed-interference configuration," *Electron. Lett.*, vol. 36, pp. 1129-1130, June 2000.

[42] J. Leuthold, B. Mikkelsen, G. Raybon, H. Joyner, J. L. Pleumeekers, B. I. Miller, K. Dreyer, and R. Behringer, "All-optical wavelength conversion between 10 and 100 Gbit/s with SOA delayed-interference configuration," *Opt. Quantum Electron.*, vol. 33, pp. 939-952, July 2001.

[43] J. Leuthold, B. Mikkelsen, R. E. Behringer, G. Raybon, C. H. Joyner, and P. A. Besse, "Novel 3R regenerator based on semiconductor optical amplifier delayed-interference configuration," *IEEE Photon. Technol. Lett.*, vol. 13, pp. 860-862, Aug. 2001.

[44] J. Leuthold, R. Ryf, D. N. Maywar, S. Cabot, J. Jaques, and S. Patel, "Nonblocking all-optical cross connect based on regenerative all-optical wavelength converter in a transparent network demonstration over 42 nodes and 16800 km," *IEEE J. Lightw. Technol.*, vol. 21, no. 11, pp. 2863-2870, Nov. 2003.

[45] J. Leuthold, D. M. Marom, S. Cabot, J. Jaques, R. Ryf, and C. R. Giles, "All-optical wavelength converter based on a pulse reformatting optical filter", Proc. Optical Fiber Communications Conference (OFC'2003), Atlanta, USA, PD41; March 2003.

[46] J. Leuthold, D. M. Marom, S. Cabot, J. Jaques, R. Ryf, and C. R. Giles, "All-optical wavelength conversion using a pulse reformatting optical filter," *IEEE J. Lightw. Technol.*, vol. 22, pp. 186-192, Jan. 2004.

[47] J. Leuthold, L. Möller, J. Jaques, S. Cabot, L. Zhang, P. Bernasconi, M. Cappuzzo, L. Gomez, E. Laskowski, E. Chen, A. Wong-Foy and A. Griffin, "160 Gbit/s SOA all-optical wavelength converter and assessment of its regenerative properties," *Electron. Lett.*, vol. 40, pp. 554-555, Apr. 2004.

[48] Z. Li, Y. Dong, J. Mo, Y. Wang, and C. Lu, "Cascaded all-optical wavelength conversion for RZ-DPSK signal based on four-wave mixing in semiconductor optical amplifier," *IEEE Photon. Technol. Lett.*, vol. 16, pp. 1685-1687, Jul. 2004.

[49] Y. Liu, E. Tangdiongga, Z. Li, S. Zhang, H. de Waardt, G. D. Khoe, and H. J. S. Dorren, "Error-free all-optical wavelength conversion at 160 Gb/s using a semiconductor optical amplifier and an optical bandpass filter," *IEEE J. Lightw. Technol.*, vol. 24, pp. 230-236, Jan. 2006.

[50] Y. Liu, E. Tangdiongga, Z. Li, H. de Waardt, A. M. J. Koonen, G. D. Khoe, X. Shu, I. Bennion, and H. J. S. Dorren, "Error-free 320-Gb/s all-optical wavelength conversion using a single semiconductor optical amplifier," *IEEE J. Lightw. Technol.*, vol. 25, pp. 103-108, Jan. 2007.

[51] C. K. Madsen and J. H. Zhao, *Optical Filter Design and Analysis*, New York: Wiley, 1999.

[52] R. J. Manning and D. A. O. Davies, "Three-wavelength device for all-optical signal processing," *Opt. Lett.*, vol. 19, pp. 889-891, 1994.

[53] R. J. Manning, X. Yang, R. P. Webb, and R. Giller, "The 'turbo-switch' – a novel technique to increase the high-speed response of SOAs for wavelength conversion," Proc. OFC'2006, Anaheim, CA, OWS8.

[54] J. Mark and J. Mørk, "Subpicosecond gain dynamics in InGaAsP optical amplifiers: Experiment and theory," *Appl. Phys. Lett.*, vol. 61 (19), pp. 2281-2283, Nov. 1992.

[55] R. McBride and J. D. C. Jones, "A passive phase recovery technique for Sagnac interferometers based on controlled loop birefringence," *IEEE J. Modem Optics*, vol. 39, pp. 1309-1320, June 1992.

[56] A. Mecozzi and J. Mørk, "Saturation induced by picosecond pulses in semiconductor optical amplifiers," *J. Opt. Soc. Am. B*, vol. 14, no. 4, pp. 761–770, April 1997.

[57] A. Mecozzi and J. Mørk, "Saturation effects in nondegenerate four-wave mixing between short optical pulses in semiconductor laser amplifiers," *IEEE J. Select. Topics Quantum Electron.*, vol. 3, pp. 1190–1207, Sept./Oct. 1997.

[58] J. Mørk, J. Mark, and C. P. Seltzer, "Carrier heating in InGaAsP laser amplifier due to two-photon absorption," *Appl. Phys. Lett.*, vol. 64, pp. 2206-2208, April 1994

[59] J. Mørk and J. Mark, "Time-resolved spectroscopy of semiconductor laser devices: experiments and modelling," *SPIE* vol. 2399, pp. 146-159, 1995.

[60] J. Mørk and A. Mecozzi, "Theory of the ultrafast optical response of active semiconductor waveguides," *J. Opt. Soc. Am. B*, vol. 13, no. 8, pp. 1803-1816, April 1996.

[61] D. B. Mortimore, "Fiber loop reflectors," *IEEE J. Lightw. Technol.*, vol. 6, pp. 1217-1224, July 1988.

[62] M. L. Nielsen, B. Lavigne and B. Dagens, "Polarity-preserving SOA-based wavelength conversion at 40 Gbit/s using bandpass filtering," *IEEE Electron. Lett.*, vol. 39, No. 18, pp. 1334-1335, Sep. 2003.

[63] M. L. Nielsen and J. Mørk, "Increasing the modulation bandwidth of semiconductor-optical-amplifier-based switches by using optical filtering ," *J. Opt. Soc. Am. B*, vol. 21, pp. 1606-1619, 2004.

[64] M. Nielsen and J. Mørk, "Bandwidth enhancement of SOA-based switches using optical filtering: theory and experimental verification," *Opt. Express*, vol. 14, pp. 1260-1265, Feb. 2006.

[65] L. Occhi, L. Schares, and G. Guekos, "Phase modeling based on the α-factor in bulk semiconductor optical amplifier," *IEEE J. Select. Topics Quantum Electron.*, vol. 9, pp. 788-797, Jan. 2003.

[66] M. Osinski and J. Buus, "Linewidth broadening factor in semiconductor lasers – an overview," *IEEE J. Quantum Electron.*, vol. QE-23, pp. 9-29, Jan. 1987.

[67] S. Radic and C. J. McKinstrie, "Optical amplification and signal processing in highly nonlinear optical fiber," *IEICE Trans. Electron.*, vol. E88-C, pp. 859-869, May 2005.

[68] T. Saitoh and T. Mukai, "Structural design for polarization-insensitive travelling-wave semiconductor laser amplifiers," *Opt. and Quantum Electron.*, vol. 21, pp. s47-s58, Jan. 1989.

[69] R. P. Schreieck, M. H. Kwakernaak, H. Jäckel, and H. Melchior, "All-optical switching at multi-100-Gb/s data rates with Mach-Zehnder interferometer switches," *IEEE J. Quantum Electron.*, vol. 38, pp. 1053–1061, Aug. 2002.

[70] S. Schuster and H. Haug, "Calculation of the gain saturation in CW semiconductor lasers with Boltzmann kinetics for Coulomb and LO phonon scattering," *Semicond. Sci. Technol.*, vol. 10, pp. 281-289, Mar. 1995.

[71] K. E. Stubkjaer, "Semiconductor optical amplifier-based all-optical gates for high-speed optical processing," *IEEE J. Select. Topics Quantum Electron.*, vol. 6, pp. 1428-1435, Nov./Dec. 2000.

[72] Y. Suematsu and A. R. Adams, *Handbook of Semiconductor Lasers and Photonic Integrated Circuits*, London, U.K.: Chapman & Hall, 1994.

[73] M. Sugawara, T. Akiyama, N. Hatori, Y. Nakata, H. Ebe, and H. Ishikawa, "Quantum-dot semiconductor optical amplifiers for high-bit-rate signal processing up to 160 Gb/s and a new scheme of 3R regenerators," *Meas. Sci. Technol.*, vol. 13, pp. 1683-1691, 2002.

[74] V. G. Ta'eed, M. R. E. Lamont, D. J. Moss, B. J. Eggleton, D. -Y. Choi, S. Madden, and B. Luther-Davies, "All optical wavelength conversion via cross phase modulation in chalcogenide glass rib waveguides," *Opt. Express*, vol. 14, pp. 11242-11247, Oct. 2006.

[75] K. Tajima, "All-optical switch with switch-off time unrestricted by carrier lifetime," *Jpn. J. Appl. Phys.*, vol. 32, pp. L1746-L1749, Dec. 1993.

[76] T. Tekin, C. Bornholdt, J. Slovak, M. Schlak, B. Sartorius, J. Kreissl, S. Bauer, C. Bobbert, W. Brinker, B. Maul, Ch. Schmidt and H. Ehlers, "Semiconductor based true all-optical synchronous modulator for 3R regeneration", in *Proc. 29th European. Conf. Opt. Commun.* (ECOC), Rimini, Italy, pp. 794-795, 2003.

[77] R. Trebino, *Frequency-Resolved Optical Gating. The Measurement of Ultrashort Laser Pulses,* Springer, 2002.

[78] Y. Ueno, S. Nakamura, K. Tajima, and S. Kitamura, "3.8-THz wavelength conversion of picosecond pulse using a semiconductor delayed-interference signal wavelength converter (DISC)," *IEEE Photon. Technol. Lett.*, vol. 10, pp 346-348, Oct. 1998.

[79] K. Vahala, L. C. Chiu, S . Margalit, and A. Yariv, "On the linewidth enhancement factor α in semiconductor injection lasers," *Appl. Phys. Lett.*, vol. 42, pp. 631-633, Apr. 1983.

[80] A. Villeneuve, M. Sundheimer, N. Finlayson, G. I. Stegemann, S. Morasca, C. Rigo, R. Calvani and C. De Bernardi, "Two-photon absorption in InGaAlAs/InP waveguides at communication wavelengths," *Appl. Phys. Lett.*, vol. 56, pp. 1865-1867, May 1990.

[81] P. Vorreau, A. Marculescu, J. Wang, G. Boettger, B. Sartorius, C. Bornholdt, J. Slovak, M. Schlak, C. Schmidt, S. Tsadka, W. Freude, J. Leuthold, "Cascadability and regenerative properties of SOA all-optical DPSK wavelength converters," *IEEE Photon. Technol. Lett.*, vol. 18, pp. 1970 - 1972, Sep. 2006.

[82] J. Wang, A. Maitra, C. G. Poulton, W. Freude, and J. Leuthold, "Temporal dynamics of the alpha factor in semiconductor optical amplifiers," *IEEE J. Lightw. Technol.*, vol. 25, pp. 891-900, Mar. 2007.

[83] W. Wang, H. N. Poulsen, L. Rau, H. Chou, J. E. Bowers and D. J. Blumenthal, "Raman-enhanced regenerative ultra-fast all-optical fiber XPM wavelength converter," *IEEE J. Lightw. Technol.*, vol. 23, pp. 1105-1115, Mar. 2005.

[84] K. J. Werner, *Rausch-und Sättigungseigenschaften von optischen 1,3-µm-Halbleiter-Wanderwellenverstärkern*, Ph.D. Thesis, Dept. Elect. Eng. and Information Eng, University of Karlsruhe, 1989.

[85] L. D. Westbrook and M. J. Adams, "Simple expressions for the linewidth enhancement factor in direct-gap semiconductors," *IEE Proceedings*, vol. 134, Pt. J, p209-214, Aug. 1987.

[86] M. Willatzen, A. Uskov, J. Mørk, H. Olesen, B. Tromborg, A. P. Jauho, "Nonlinear gain suppression in semiconductor lasers due to carrier heating," *IEEE Photon. Technol. Lett.*, vol. 3, pp. 606-609, July 1991.

[87] L. Xu, I. Glesk, V. Baby, and P. R. Prucnal, "All-optical wavelength conversion using SOA at nearly symmetric position in a fiber-based Sagnac interferometer loop," IEEE Photon Technol. Lett., vol. 16, pp. 539-541, Feb. 2004.

[88] L. Xu, V. Baby, I. Glesk, P. R. Prucnal, "New description of transmission of an SOA based Sagnac loop and its application for NRZ wavelength conversion", *Opt. Commun.*, vol. 244, pp. 199-208, Jan. 2005.

[89] A. Yariv, *Optical electronics in modern communications* 5th Ed., Oxford University Press, Inc., 1997.

[90] S. J. B. Yoo, "Wavelength conversion technologies for WDM network applications," *IEEE J. Lightw. Technol.*, vol. 14, pp. 955-966, June 1996.

[91] H. -Y. Yu, D. Mahgerefteh, P. S. Cho, and J. Goldhar, "Optimization of the frequency response of a semiconductor optical amplifier wavelength converter using a fiber Bragg grating," *IEEE J. Lightw. Technol.*, vol. 17, pp. 308-315, Feb. 1999.

[92] L. Zhang, I. Kang, A. Bhardwaj, N. Sauer, S. Cabot, J. Jaques, D.T. Nielson, "Reduced recovery time semiconductor optical amplifier using p-type-doped multiple quantum wells, " *IEEE Photon. Technol. Lett.*, vol. 18, pp. 2323-2325, Nov. 2006.

[93] D. Zwillinger, *Handbook of differential equations*, Academic Press, 2nd Ed., 1992.

[94] RSoft design group, Inc., "OptSim 4.6 user guide," website: http://www.rsoftdesign.com/

[95] *Ultrashort Laser Pulses I*, script, Georgia Institute of Technology. [Online] http://www.physics.gatech.edu/gcuo/lectures/UFO03UltrashortpulsesI.ppthttp://www.ph ysics.gatech.edu/gcuo/lectures/UFO03Ultrashort pulsesI.ppt.

Appendix A   Nonlinear Gain Compression

The application of semiconductor optical amplifier in all-optical network should combine a high gain but large and ultrafast material nonlinearities. In the work of Mecozzi and Mørk in [56] and [57], the dynamics of carrier heating (CH) and spectral-hole burning (SHB) are taken into account in the evolution of the gain saturation.

Following [56], [57] and [59], the rate equations for the local carrier densities n_β ($\beta = c, v$ for conduction band (CB) and valence band (VB), respectively), carrier density N, and carrier temperatures T_β of the active region are:

$$\frac{\partial n_\beta}{\partial t} = -\frac{n_\beta - \overline{n}_\beta}{\tau_{1\beta}} - v_g\, g_m\, S\,, \tag{A.1}$$

$$\frac{\partial n_\beta}{\partial t} = -\frac{n_\beta - \overline{n}_\beta}{\tau_{1\beta}} - v_g\, g_m\, S\,, \tag{A.2}$$

$$\frac{\partial T_\beta}{\partial t} = -\frac{T_\beta - T_L}{\tau_{h,\beta}} + \left(\frac{\partial U_\beta}{\partial T_\beta}\right)_N^{-1} \left[\frac{\sigma_\beta N h f_0}{g_m} + \left(\frac{\partial U_\beta}{\partial N}\right)_{T_\beta} - W_{\beta,0}\right] v_g\, g_m\, S\,. \tag{A.3}$$

The parameters are: Fermi densities $\overline{n}_\beta$, group velocity v_g, material gain g_m, photon density S, intraband scattering times $\tau_{1\beta}$, temperature relaxation times $\tau_{h\beta}$, carrier lifetime τ_s, current I, elementary charge e, active region volume V, energy density U_β, free-carrier absorption coefficient σ_β, Planck constant h, optical frequency f_0, transition energy $W_{\beta,0}$, lattice temperature T_L. The Fermi densities $\overline{n}_\beta$ appeared in (A.1) are defined through the Fermi function f evaluated at the instantaneous value of carrier density and temperature:

$$\overline{n}_\beta = N_0 f(W_{\beta,0}; N, T_\beta)\,, \tag{A.4}$$

where N_0 is the density of available states in the optically coupled region.

For pulse widths much longer than the scattering times $\tau_{1,\beta}$ and $\tau_{h,\beta}$, i.e., several picoseconds long, the rate equations for the local carrier densities (A.1) can be solved under the adiabatic limit, which means the dephasing time of the dipole is sufficiently shorter than the other characteristic times,

$$n_\beta = \overline{n}_\beta - \tau_{1\beta} v_g\, g_m\, S\,, \tag{A.5}$$

where $\overline{n}_\beta$ can be separated into the equilibrium term for given carrier density at the lattice temperature and the contribution due to the carrier temperature change:

$$\overline{n}_\beta = \overline{n}_{\beta,T_L} + \left(\frac{\partial \overline{n}_\beta}{\partial T_\beta}\right)_N \Delta T_\beta\,. \tag{A.6}$$

The temperature change $\Delta T_\beta = T_\beta - T_L$ obeys also the rate equation (A.3)

$$\frac{\partial \Delta T_\beta}{\partial t} = -\frac{\Delta T_\beta}{\tau_{h,\beta}} + K_\beta v_g g_m S , \tag{A.7}$$

with

$$K_\beta = \left(\frac{\partial U_\beta}{\partial T_\beta}\right)_N^{-1} \left[\left(\frac{\partial U_\beta}{\partial N}\right)_{T_\beta} - W_{\beta,0}\right], \tag{A.8}$$

where the free-carrier absorption coefficient σ_β is set to be 0.

Following [57], the material gain coefficient g_m for non-equilibrium situation can be written as

$$g_m(N,\lambda,T) = \bar{g}_m(N,\lambda,T_L) + \beta = c, v \sum (\Delta g_{\beta,h} + \Delta g_\beta) , \tag{A.9}$$

$$\bar{g}_m(N,\lambda,T_L) = \frac{a_N}{v_g}(\bar{n}_{c,T_L} + \bar{n}_{v,T_L} - N_0) = a_0(N - N_{tr}), \tag{A.10}$$

$$\Delta g_{\beta,h} = \frac{a_N}{v_g} \Delta n_{\beta,h} = \frac{a_N}{v_g}(\bar{n}_\beta - \bar{n}_{\beta,T_L}), \tag{A.11}$$

$$\Delta g_\beta = \frac{a_N}{v_g} \Delta n_\beta = \frac{a_N}{v_g}(n_\beta - \bar{n}_\beta). \tag{A.12}$$

Here a_N/v_g is the gain cross section and a_0 is the differential gain and N_{tr} is the carrier density at transparency. Practically, the material gain coefficient $\bar{g}_m$ at quasi-equilibrium can be evaluated at given carrier density N and a wavelength λ with an appropriate model [40]. Then, the dynamics of the coefficient $\bar{g}_m$ is determined via (A.2). Other contributions to gain saturation are given as

$$\frac{\partial \Delta g_\beta}{\partial t} = -\frac{\Delta g_\beta}{\tau_{1\beta}} - \frac{\varepsilon_{\mathrm{SHB},\beta}}{\tau_{1\beta}} g_m S(t,z) - \left(\frac{\partial \Delta g_{\beta,h}}{\partial t} + y_\beta \frac{\partial \bar{g}_m}{\partial t}\right), \tag{A.13}$$

$$\frac{\partial \Delta g_\beta,h}{\partial t} = -\frac{\Delta g_\beta,h}{\tau_h\beta} - \frac{\varepsilon_{\mathrm{CH},\beta}}{\tau_h\beta} g_m S(t,z) , \tag{A.14}$$

The nonlinear gain suppression factors due to carrier heating (CH) and spectral hole burning (SHB) are defined as

$$\varepsilon_{\mathrm{CH},\beta} = -a_N K_\beta \tau_{h\beta} \frac{\partial n_\beta}{\partial T_\beta} = -v_g K_\beta \tau_{h\beta} \frac{\partial g_m}{\partial T_\beta} , \tag{A.15}$$

$$\varepsilon_{\mathrm{SHB},\beta} = a_N \tau_{1\beta} , \tag{A.16}$$

with

$$y_\beta = \frac{a_N}{a_0 v_g} \frac{\partial \bar{n}_{\beta,l}}{\partial N} , \tag{A.17}$$

and $\beta \sum y_\beta = 1$.

The local carrier densities $\bar{n}_{\beta,T_L}$ are the quasi-equilibrium densities evaluated at the lattice temperature. Following [57] they can be calculated with Fermi's golden rule as

$$\bar{n}_{c,T_L} = \int_{W_c}^{\infty} f_c(W)\rho_c(W)dW , \tag{A.18}$$

$$\bar{n}_{v,T_L} = \int_{-\infty}^{W_v} f_v(W)\rho_v(W)dW , \tag{A.19}$$

where $\rho_\beta(W)$ are the densities of states and $f_\beta(W)$ are the occupation probabilities in CB and VB. W_c and W_v are the bandedges of the CB and VB. With the assumption of a non-degenerate semiconductor (even it is not a usual case for practical SOAs) and the simplification of the parabolic band structure, $\bar{n}_{c,T_L}$ and $\bar{n}_{v,T_L}$ become

$$\bar{n}_{c,T_L} = N_c \exp(-\frac{W_c - W_{f_c}}{k_B T_L}) , \tag{A.20}$$

$$\bar{n}_{v,T_L} = P_c \exp(-\frac{W_{f_v} - W_v}{k_B T_L}) , \tag{A.21}$$

where k_B is the Boltzmann constant. W_{f_c} and W_{f_v} are the quasi-Fermi levels. N_c and P_c are the effective densities of states in the CB and VB respectively, and given by

$$N_c = 2\left(\frac{m_e k_B T_L}{2\pi\hbar^2}\right)^{3/2} ; \quad P_c = 2\left(\frac{m_v k_B T_L}{2\pi\hbar^2}\right)^{3/2} , \tag{A.22}$$

where m_e is the effective mass of electron in the CB and m_v is the effective mass of hole in the VB. m_v can be evaluated from the effective mass of a heavy hole m_{hh} and the effective mass of a light hole m_{lh} in the VB,

Reference [56] also shows that Eq. (A.13) and (A.14) can reduce to the well-known expression for the amplifier gain for pulses much longer that the intraband scattering times $\tau_{1\beta}$ and $\tau_{h\beta}$. Under this condition, Δn_β and $\Delta n_{\beta,h}$, and hence Δg_β and $\Delta g_{\beta,h}$, can be assumed in quasi-equilibrium with the instantaneous value of the field. Therefore, the derivatives in Eq. (A.13) and (A.14) can be neglected. This assumption then leads to

$$\Delta g_{\beta,h} = -\varepsilon_{\mathrm{SHB},\beta}\, g_m\, S(t,z) , \tag{A.23}$$

$$\Delta g_{\beta,h} = -\varepsilon_{\mathrm{SHB},\beta}\, g_m\, S(t,z) . \tag{A.24}$$

Thus the material gain coefficient g_m in the quasi-equilibrium is

$$g_m = \frac{\bar{g}_m}{1 + \varepsilon_{\mathrm{tot}}\, S(t,z)} , \tag{A.25}$$

with

$$\varepsilon_{\text{tot}} = \beta = c, v \sum (\varepsilon_{\text{CH},\beta} + \varepsilon_{\text{SHB},\beta}) .$$

(A.26)

The Eq. (A.2), (A.14) and (A.23) are implemented in Chapter 2 to model the dynamics of the material gain in an SOA.

Appendix B   Parameters used in SOA Modeling

Table B. 1. Parameters for material gain parameterization, derived from [39]. The value marked with (*) is for an SOA with a length of 1.6 mm

Symbol	Parameter	Value	Unit
N_0	Transparency carrier density	$7.8 \cdot 10^{23}$	m^{-3}
a_0	Differential gain constant	$5.0 \cdot 10^{-20}$	m^2
a_1	Fitting parameter	1.2	–
$\lambda_{p,0}$	Peak wavelength λ_p at N_0	$1.585 \backslash 1608$ *	μm
b_0	Derivative of λ_p with respect to N	$6.85 \cdot 10^{-26}$	$m^3 \, \mu m$
b_1	Fitting parameter	$-3.5 \cdot 10^{-51}$	$m^6 \, \mu m$
$\lambda_{z,0}$	Bandgap wavelength at N_0	1.670	μm
z_0	Derivative of λ_z with respect to N	$-1.5 \cdot 10^{-27}$	$m^3 \, \mu m$

Table B. 2. Parameters for α_N parameterization derived from the measurement results in [39].

Symbol	Parameter	Value	Unit
$\lambda_{\alpha,0}$	Wavelength fitting parameter in α_N	1.549	μm
c_α	Derivative of λ_α with respect to N	-9.31×10^{-26}	$m^3 \, \mu m$
α_{N_0}	α_N at N_0 and $\lambda_{\alpha 0}$	3.8	–
α_{N_1}	Dependence of α_N on N and λ	18	–

Table B. 3. Parameters for internal loss parameterization derived from the measurement results in [39]. The value marked with (*) is for an SOA with a length of 1.6 mm.

Symbol	Parameter	Value	Unit
$\alpha_{int,0}$	Total internal loss at $\lambda_{p,0}$	$7000 \backslash 9500$ *	m^{-1}
K_1	Derivative of α_{int} with respect to λ	-8100	$\left(m \cdot \mu m \right)^{-1}$

Table B. 4. Parameters for confinement factor parameterization, derived from the measured results in [39].

Symbol	Parameter	Value	Unit
N_{Γ_0}	Fitting carrier density	$2.2 \cdot 10^{24}$	m^{-3}
λ_{Γ_0}	Fitting wavelength	1550	μm
Γ_0	Γ at λ_{Γ_0} and N_{Γ_0}	0.35	–
$\left.\dfrac{\partial\Gamma}{\partial\lambda}\right|_{N_{\Gamma_0},\lambda_{\Gamma_0}}$	Derivative of Γ with respect to λ	$-4.166 \cdot 10^{-4}$	nm^{-1}
$\left.\dfrac{\partial\Gamma}{\partial N}\right|_{N_{\Gamma_0},\lambda_{\Gamma_0}}$	Derivative of Γ with respect to N	$-5.7 \cdot 10^{-27}$	m^3

Table B. 5. Other SOA parameters used in the calculation derived from the cross-gain experiment results.

Symbol	Parameter	Value	Unit
σ	Active cross section	0.225	μm^2
L	SOA Length	$2.6 \backslash 1.6*$	mm
n_g	Effective group refractive index	3.57	–
A_{SRH}	Nonradiative recombination	$13.5 \cdot 10^8$	s^{-1}
B_{BB}	Bimolecular recombination	$5.6 \cdot 10^{-16}$	$m^3 s^{-1}$
C_{Aug}	Auger recombination	$1.5 \cdot 10^{-40}$	$m^6 s^{-1}$
D_{leak}	Leakage recombination	$5 \cdot 10^{-100}$	$m^{13.5} s^{-1}$
β_2	TPA coefficient	$4 \cdot 10^{-21}$	m^2
$\tau_{T,c}$	CH time constant in CB	700	fs
$\tau_{T,v}$	CH time constant in VB	500	fs
$\varepsilon_{CH,c}$	CH gain suppression factor in CB	$1.45 \cdot 10^{-23}$	m^3
$\varepsilon_{CH,v}$	CH gain suppression factor in VB	$1.03 \cdot 10^{-23}$	m^3
$\varepsilon_{SHB,c}$	SHB gain suppression factor in CB	$2.11 \cdot 10^{-23}$	m^3
$\varepsilon_{SHB,v}$	SHB gain suppression factor in VB	$1.50 \cdot 10^{-23}$	m^3
α_T	α-factor for CH	0.7	-
α_{TPA}	α-factor for TPA	-3	-
R_1 & R_2	Facet reflectivities	10^{-4}	-

Appendix C   Predictor-Corrector Method

Because the differential equations for carrier density and gain compression terms, Eq. (2.23) and (2.33), are stiff, a predictor-corrector method is implemented to obtain an implicit solution. This method is following reference [13]

Problem:

$$\frac{dy}{dt} = f[t, x(t), y(t)], \quad f_y(t, x, y) = \frac{\partial f(t, x, y)}{\partial y} < 0. \tag{C.1}$$

All the variables sampled at the time step t_i are x_i and y_i.

Predictor for $(i+1)-$ step calculation with a step size $\Delta t = t_{i+1} - t_i$:

$$y_{i+1}^{(0)} = y_i + \Delta t \, f(t_i, x_i, y_i). \tag{C.2}$$

$(v+1)-$ Corrector:

$$y_{i+1}^{(v+1)} = y_i + \Delta t \, f(t_{i+1}, x_{i+1}, y_{i+1}^{(v+1)}) = y_{i+1}^{(v)} + \Delta y_{i+1}^{(v+1)}, \quad \text{for} \quad v \geq 0, \tag{C.3}$$

With

$$f(t_{i+1}, x_{i+1}, y_{i+1}^{(v+1)}) \approx f(t_{i+1}, x_{i+1}, y_{i+1}^{(v)}) + \Delta y_{i+1}^{(v+1)} f_y(t_{i+1}, x_{i+1}, y_{i+1}^{(v)}), \tag{C.4}$$

$$f_y(t_{i+1}, x_{i+1}, y_{i+1}^{(v)}) = \frac{f(t_{i+1}, x_{i+1}, y_{i+1}^{(v)}) - f(t_{i+1}, x_{i+1}, y_{i+1}^{(v-1)})}{y_{i+1}^{(v)} - y_{i+1}^{(v-1)}}, \tag{C.5}$$

$$f(t_{i+1}, x_{i+1}, y_{i+1}^{(-1)}) = f(t_i, x_i, y_i), \quad y_{i+1}^{(-1)} = y_i. \tag{C.6}$$

So

$$\Delta y_{i+1}^{(v+1)} = \frac{y_i - y_{i+1}^{(v)} + \Delta t \, f(t_{i+1}, x_{i+1}, y_{i+1}^{(v)})}{1 - \Delta t \, f_y(t_{i+1}, x_{i+1}, y_{i+1}^{(v)})}, \tag{C.7}$$

$$y_{i+1}^{(v+1)} = y_{i+1}^{(v)} + \Delta y_{i+1}^{(v+1)} = y_{i+1}^{(v)} + \frac{y_i - y_{i+1}^{(v)} + \Delta t \, f(t_{i+1}, x_{i+1}, y_{i+1}^{(v)})}{1 - \Delta t \, f_y(t_{i+1}, x_{i+1}, y_{i+1}^{(v)})}. \tag{C.8}$$

End of Iteration:

$$|\Delta y_{i+1}^{(v+1)}| \leq \varepsilon. \tag{C.9}$$

Result:

$$y_{i+1} = y_{i+1}^{(v+1)}, \quad \text{with residual error } \varepsilon = O(\Delta t). \tag{C.10}$$

We can also take the average between predictor $y_{i+1}^{(0)}$ and the corrector after iteration $y_{i+1}^{(v+1)}$,

$$y_{i+1} = \frac{y_{i+1}^{(v+1)} + y_{i+1}^{(0)}}{2}. \tag{C.11}$$

This is equivalent to the improved Euler's method (or 2nd Runge-Kutta method [93])

$$y_{i+1}^{*} = y_i + h f(t_i, x_i, y_i),$$
(C.12)

$$y_{i+1} = y_i + \frac{h}{2}[f(t_i, x_i, y_i) + f(t_{i+1}, x_{i+1}, y_{i+1}^{*})],$$
(C.13)

which is of second order accuracy with a rest error $\varepsilon = O((\Delta t)^2)$.

Example:

$$\text{Problem}: \frac{dy}{dt} = f(t, y) = -\frac{1}{2} y^3 \quad \text{with an exact solution}: y = t^{-1/2}.$$
(C.14)

$$\text{Predictor}: y_{i+1}^{(0)} = y_i + \Delta t(-\frac{1}{2} y_i^3).$$
(C.15)

$$(v+1) - \text{Corrector}: y_{i+1}^{(v+1)} = y_{i+1}^{(v)} + \frac{y_i - y_{i+1}^{(v)} - \frac{\Delta t}{2}(y_{i+1}^{(v)})^3}{1 - \Delta t f_y(t_{i+1}, y_{i+1}^{(v)})},$$
(C.16)

$$\text{with } f_y(t_{i+1}, y_{i+1}^{(v)}) = \frac{-\frac{1}{2}(y_{i+1}^{(v)})^3 + \frac{1}{2}(y_{i+1}^{(v-1)})^3}{y_{i+1}^{(v)} - y_{i+1}^{(v-1)}}.$$
(C.17)

$$\text{Result}: y_{i+1} = y_{i+1}^{(v+1)}, \quad \text{for } |\Delta y_{i+1}^{(v+1)}| \leq \varepsilon$$
(C.18)

Appendix D Fiber Loop Reflector

Calculation of Jones Matrix for a Waveplate

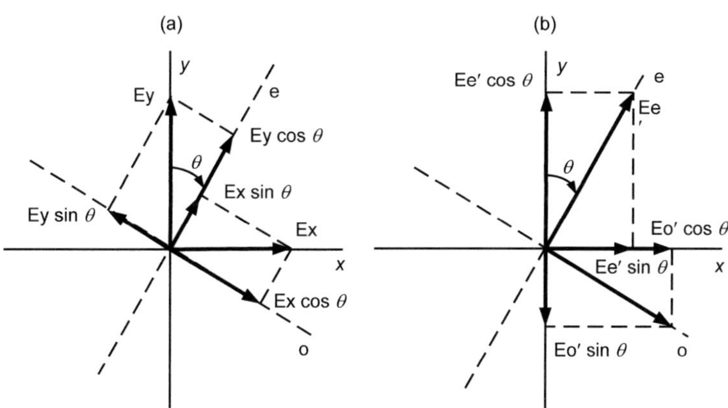

Fig. D. 1 Calculation of Jones matrix coefficients.

Calculation of Jones matrix for a waveplate having the retardation ϕ and orientation θ can be explained by using Fig. D. 1. Orthogonal field components E_x and E_y imping on a birefringent medium having fast and slow axes (e and o) orientated at an angle θ with respect to the y and x axes, shown in Fig. D. 1(a). The components of field along the e and o axes are then:

$$E_e = E_x \sin\theta + E_y \cos\theta, \tag{D.1}$$

$$E_o = E_x \cos\theta - E_y \sin\theta. \tag{D.2}$$

These fields propagate through the medium with different velocities and possibly different attenuation. If we assume the o component field has a zero relative phase shift with respect to the phase lead of the e component field, then the exit fields E_e' and E_o', Fig. D. 1(b), are given by

$$E_e' = E_e T_e \exp(j\phi); \qquad E_o' = E_o T_o, \tag{D.3}$$

where the retardation ϕ is the phase lead of the field component on the fast axis e with respect to the field component on the slow axis o of the birefringent medium (in a so-called half waveplate, the phase lead is $180°$).

T_e and T_o are the amplitude transmission coefficients for the e and o components. The output fields E_x' and E_y' are therefore given by:

$$E_x' = E_o' \cos\theta + E_e' \sin\theta, \tag{D.4}$$

$$E_y' = E_e' \cos\theta - E_o' \sin\theta. \tag{D.5}$$

Substituting (1), (2) and (3) into (4) and (5) we can obtain

$$E_x' = [T_e \exp(j\phi)\sin^2\theta + T_o \cos^2\theta]E_x + [T_e \exp(j\phi)\cos\theta\sin\theta - T_o \sin\theta\cos\theta]E_y, \tag{D.6}$$

$$E_y' = [T_e \exp(j\phi)\sin\theta\cos\theta - T_o \sin\theta\cos\theta]E_x + [T_e \exp(j\phi)\cos^2\theta - T_o \sin^2\theta]E_y. \tag{D.7}$$

By comparing these equations with

$$\begin{pmatrix} E_x' \\ E_y' \end{pmatrix} = \begin{pmatrix} J_{xx} & J_{xy} \\ J_{yx} & J_{yy} \end{pmatrix}\begin{pmatrix} E_x \\ E_y \end{pmatrix}, \tag{D.8}$$

where the Jones matrix elements can be determined as

$$J_{xx} = T_e \exp(j\phi)\sin^2\theta + T_o \cos^2\theta, \tag{D.9}$$

$$J_{xy} = J_{yx} = [T_e \exp(j\phi) - T_o]\cos\theta\sin\theta, \tag{D.10}$$

$$J_{yy} = T_e \exp(j\phi)\cos^2\theta + T_o \sin^2\theta. \tag{D.11}$$

For $T_e = T_o = 1$, there are no attenuation for the e and o components. In this lossless case, the elements of the Jones matrix fulfill the unitary condition

$$|J_{xx}|^2 + |J_{yx}|^2 = 1, |J_{xy}|^2 + |J_{yy}|^2 = 1, J_{xx}J_{xy}^* + J_{yx}J_{yy}^* = 0. \tag{D.12}$$

Propagation of a Field through a Fiber Loop Reflector with a Birefringent Device

We consider the effects of birefringence in the fiber loop, Fig. D. 2. For clarity of explanation it is assumed in the following discussion that the birefringence occurs near the coupler, between (f) and (c) and on the right side of Fig. D. 2. Birefringence causes a variation of optical path length with polarization angle. For example, light incident polarized along the fast axis of a discrete half wave plate will suffer a π phase difference with respect to light incident along the orthogonal slow axis. If this discrete half wave plate (or a fiber loop polarization controller set as a half wave plate) were placed in the loop, with its fast axis at θ, then light incident to the loop at an angle η will produce two counter propagating fields incident to the waveplate at angles of $+\eta$ and $-\eta$. These two field vectors are then both rotated in the same direction such that on exit from the waveplate their field vectors both point inside the loop at angles of $-(\pi - 2\theta + \eta)$ and $-(\pi - 2\theta - \eta)$ as shown by the vectors of (c) and (f). Note that from b→f' and c→g, there exists a π change of observation angle. When these fields reenter the coupler, there is no phase shift for f→i_f and there is a π phase shift for g→i_g, since a→c' and g→i_g passes the coupler twice. Meanwhile, from f→h_f and g→h_g both have a $\pi/2$ phase shift. Now, between vectors i_f and i_g, there is a relative phase difference of

$\pi - 4\theta$, while there is a relative phase difference of 4θ between vectors h_f and h_g. For instance, if the orientation θ of the half wave plate with respect to the y-axis is $\pi/4$, it would be a fiber loop mirror and the reflectivity is zero.

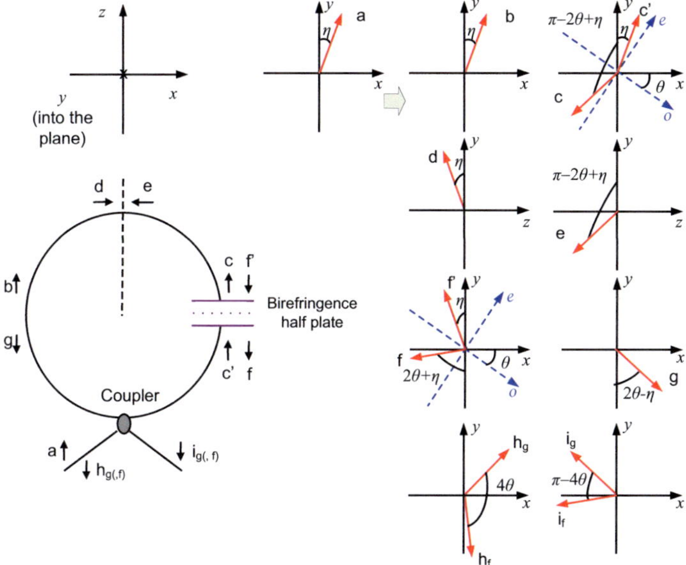

Fig. D. 2 Propagation of an instantaneous field vector around the fiber loop.

Appendix E   Approximation of the Output Phase after an SOA

In this appendix, the approximation of the phase of SOA's output is given from [19]. The notations of the variables in [19] have been adapted according to those used in this dissertation.

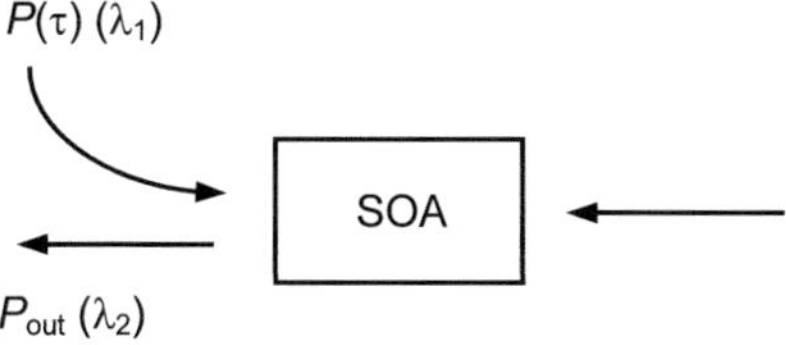

Fig. E. 1 Schematic representation of the SOA in a cross-gain–phase operation.

The general configuration of a SOA wavelength converter is presented in Fig. E. 1. The incident signal modulated on the carrier wavelength λ_1 and a cw beam at wavelength λ_2 enter the SOA from opposite directions (see Fig. E. 1). The SOA dynamics can be described by three equations: the spatial derivative (with respect to the longitudinal coordinate z) of the optical power P, its phase φ and the temporal derivative of the gain coefficient g:

$$\frac{\partial P}{\partial z} = (g - \alpha_{\text{int}})P, \tag{E.1}$$

$$\frac{\partial \varphi}{\partial z} = -\frac{1}{2}\alpha_H g, \tag{E.2}$$

$$\frac{\partial g}{\partial \tau} = \frac{g_0 - g}{\tau_c} - \frac{gP}{E_s}, \tag{E.3}$$

where E_s is the saturation energy, α_{int} is the distributed loss (it will be ignored in the following discussion, but this conjecture is not a restrictive one), P is the light power according to $P(z = 0, \tau) = P_{\text{in}}(\tau)$ and $P(z = L, \tau) = P_{\text{out}}(\tau)$, L is the SOA's length, φ is the phase, α_H is the linewidth enhancement factor, and τ_c is the spontaneous carrier lifetime.

The solution to the set of equations (E.1)-(E.3) of the converted beam (and only at the converted wavelength) is

$$P_{\text{out}}(\tau) = P_0 \exp[h(\tau)], \tag{E.4}$$

$$\varphi_{\text{out}}(\tau) = \varphi_{\text{in}}(\tau) - \frac{1}{2}\alpha_H h(\tau), \tag{E.5}$$

where $h(\tau) = \int_0^L g(z,\tau)dz$ solves the ordinary differential equation

$$\frac{dh}{d\tau} = \frac{g_0 L - h}{\tau_c} - \frac{P_{in}(\tau)}{E_s}[\exp(h) - 1].$$ (E.6)

Note that the function $h(\tau)$ is different to that used in Chapter 4. Assume that the SOA is short enough (shorter than the length that a single bit transits) and $P_{in} = P_0 + p(\tau)$, where P_0 is the input cw power at the conversion wavelength and p is the modulation power for the input wavelength. When the SOA operates in the weak modulation regime, $p(\tau) << P_0$ for every τ. Also, it is assumed that $h(\tau) = h_0 + \delta h(\tau)$, where $\delta h_0 << h_0$ for every τ.

Therefore h_0 solves the transcendental equation $g_0 L - h_0 = \Phi[\exp(h_0) - 1]$, where $\Phi \equiv P_0 \tau_c / E_s$ and δh satisfies

$$\frac{d\delta h}{d\tau} + \frac{\delta h}{\tau_e} = -p(\tau)C,$$ (E.7)

where $1/\tau_e \equiv 1/\tau_c + (P_0/E_s)\exp(h_0)$ is the reciprocal of the effective relaxation time and $C \equiv [\exp(h_0) - 1]/E_s$ is a constant that depends on the working-point parameters. The solution to Eq. (E.7) is straightforward:

$$\delta h = -C \exp(-\tau/\tau_e) \int_{-\infty}^{\tau} p(\xi) \exp(\xi/\tau_e) d\xi.$$ (E.8)

For a relatively long relaxation time this integral can be simplified to

$$\delta h \cong -C[\tau_e p_{dc} + \int_{-\infty}^{\tau} p_{ac}(\xi) d\xi],$$ (E.9)

where $p(\xi) = p_{dc} + p_{ac}(\xi)$ is separated into its dc and ac parts. Thus in the first approximation the phase of the SOA's output is proportional to an integration over the ac component of the incident signal p_{ac}:

$$\varphi_{out}(\tau) = \varphi_{in}(\tau) + \frac{1}{2}\alpha_H C[\tau_e p_{dc} + \int_{-\infty}^{\tau} p_{ac}(\xi) d\xi].$$ (E.10)

List of Publications

Journals

1. **J. Wang**, A. Marculescu, J. Li, P. Vorreau, S. Tzadok, S. Ben Ezra, S. Tsadka, W. Freude, J. Leuthold, "Pattern effect removal technique for semiconductor optical amplifier-based wavelength conversion," *IEEE Photon. Technol. Lett.*, vol. 19, no. 24, pp. 1955-1957, Dec. 2007.

2. **J. Wang**, A. Maitra, C. G. Poulton, W. Freude, J. Leuthold, "Temporal dynamics of the alpha factor in semiconductor optical amplifiers," *IEEE J. Lightw. Technol.*, vol. 25, no. 3, pp. 891-900, Mar. 2007.

3. P. Vorreau, A. Marculescu, **J. Wang**, G. Boettger, B. Sartorius, C. Bornholdt, J. Slovak, M. Schlak, C. Schmidt, S. Tsadka, W. Freude, J. Leuthold, "Cascadability and regenerative properties of SOA all-optical DPSK wavelength converters," *IEEE Photon. Technol. Lett.*, vol. 18, no. 18, pp. 1970 - 1972, Sep. 2006.

4. A. Maitra, C. G. Poulton, **J. Wang**, J. Leuthold, W. Freude, "Low switching threshold using nonlinearities in stopband-tapered waveguide Bragg gratings," *IEEE J. of Quantum Electron.*, vol. 41, no. 7, pp. 1303-1308, Oct. 2005.

5. **J. Wang**, A. Maitra, W. Freude, J. Leuthold, "Regeneration and noise suppression in XPM and XGM based DPSK wavelength conversion," in preparation.

6. **J. Wang**, A. Maitra, A. Marculescu, W. Freude, J. Leuthold, "Investigation of NRZ wavelength conversion with SOA-based Sagnac loop above 40 Gb/s," in preparation.

Conference Proceedings

1. **J. Wang**, A. Marculescu, J. Li, Z. Zhang, W. Freude, J. Leuthold, "All-optical vestigial-sideband signal generation and pattern effect mitigation with a SOA based red-shift optical filter wavelength converter", Proc. 34th European Conf. Opt. Commun. (ECOC'08), paper We.2.C.6, Brussels, Belgium, Sept. 21-25, 2008.

2. R. Bonk, P. Vorreau, S. Sygletos, T. Vallaitis, **J. Wang**, W. Freude, J. Leuthold, R. Brenot, G. H. Duan C. Meuer, S. Liebig, M. Laemmlin, D. Bimberg, "Performing cross-gain modulation with improved signal quality in an interferometric configuration", Proc. Opt. Fiber Communication Conf. (OFC'08), San Diego, USA, JWA70, Feb. 2008.

3. **J. Wang**, A. Maitra, W. Freude, J. Leuthold, "Ultrafast optical processing with chirping semiconductor optical amplifiers", Asia-Pacific Optical Communications (APOC'07), APOC02, pp. 6782-6, Wuhan, China, Nov. 1-5, 2007.

4. A. Marculescu, **J. Wang**, J. Li, P. Vorreau, S. Tzadok, S. Ben Ezra, S. Tsadka, W. Freude, J. Leuthold, "Pattern effect removal technique for semiconductor optical amplifier-based wavelength conversion", Proc. 33th European Conf. Opt. Commun. (ECOC'07), paper Tu3.4.6, Berlin, Germany, Sept. 16-20, 2007.

5. **J. Wang**, A. Maitra, W. Freude, J. Leuthold, "Regenerative properties of interferometric cross-gain and cross-phase modulation DPSK wavelength converters", Proc. Nonlinear Photonics Topical Meeting and Tabletop Exhibit 2007, paper NTuB2, Quebec City, Canada, Sept. 2-6, 2007.

6. A. Maitra, **J. Wang**, W. Freude, J. Leuthold, "High extinction ratio switching using two-photon absorption in a silicon waveguide resonator", Proc. Nonlinear Photonics Topical Meeting and Tabletop Exhibit 2007, paper JWA9, Quebec City, Canada, Sept. 2-6, 2007.

7. **J. Wang**, Y. Jiao, R. Bonk, W. Freude, J. Leuthold, "Regenerative properties of bulk and quantum dot SOA based all-optical Mach-Zehnder interferometer DPSK wavelength converters", Proc. International Conference on Photonics in Switching 2006, Herakleion (Crete), Greece, 16-18 Oct. 2006.

8. **J. Wang**, C. G. Poulton, A. Maitra, S. Cabot, J. Jaques, W. Freude, J. Leuthold, "Dynamics of linewidth-enhancement factor in semiconductor optical amplifiers", Proc. Digest OSA Topical Meeting on Optical Amplifiers and Their Applications (OAA'06), paper OTuC3, Whistler, Canada, June 25-28, 2006.

9. J. F. Pina, H. J. A. da Silva, P. N. Monteiro, **J. Wang**, W. Freude, J. Leuthold, "Cross-gain modulation-based 2R regenerator using quantum-dot semiconductor optical amplifiers at 160 Gbit/s", Proc. 9th Intern. Conf. on Transparent Optical Networks (ICTON'07), paper Tu.A1.8, pp. 106-109, Rome, Italy, July 1-5, 2007,

10. J. Leuthold, **J. Wang**, T. Vallaitis, Ch. Koos, R. Bonk, A. Marculescu, P. Vorreau, S. Sygletos, W. Freude, "New approaches to perform all-optical signal regeneration", Proc. 9th Intern. Conf. on Transparent Optical Networks (ICTON'07), paper We.D2.1, pp. 222-225 (invited), Rome, Italy, July 1-5, 2007.

11. G. Böttger, J.-M. Brosi, A. Maitra, **J. Wang**, A. Y. Petrov, M. Eich, J. Leuthold, W. Freude, "Broadband slow light and nonlinear switching devices", Proc. Progress in Electromagnetics Research Symposium 2007 (PIERS'07), pp. 1338-1342 (invited), Beijing, China, March 26-30, 2007.

12. A. Maitra, **J. Wang**, J. Leuthold, W. Freude "Isolator behaviour in asymmetric $\pi/4$ phase-shifted structure", Proc. 8th Intern. Conf. on Optoelectronics, Fiber Optics & Photonics (Photonics'06), paper IOD17, Hyderabad, India, Dec. 13-16, 2006.

13. W. Freude, J. Leuthold, P. Vorreau, A. Marculescu, **J. Wang**, G. Böttger, "All-Optical processing of novel modulation formats using semiconductor optical amplifiers", Proc. Optical Society of America Annual Meeting (OSA'06), Rochester (NY), USA, Oct. 08-12, 2006 (invited tutorial).

14. J. Leuthold, W. Freude, P. Vorreau, A. Marculescu, **J. Wang**, G. Böttger, "All-optical signal processing for phase-sensitive modulation formats", Proc. SPIE, vol.6353, Asia-Pacific Optical Communications (APOC'06), Gwangju, Korea, paper 635311, APOC 03, Sept. 2006. (invited tutorial)

15. J. Leuthold, W. Freude, G. Böttger, **J. Wang**, A. Marculescu, P. Vorreau, R. Bonk, "All-optical regeneration", Proc. 8th Intern. Conf. on Transparent Optical Networks (ICTON'06), Nottingham, UK, June 18-22, 2006 (invited).

16. W. Freude, A. Maitra, **J. Wang**, Ch. Koos, C. G. Poulton, M. Fujii, J. Leuthold, "All-optical signal processing with nonlinear resonant devices", Proc. 8th Intern. Conf. on Transparent Optical Networks (ICTON'06), Nottingham, UK, June 18-22, 2006 (invited).

17. J. Leuthold, W. Freude, G. Boettger, P. Vorreau, A. Marculescu, **J. Wang**, "Transparent networks and the role of SOA based all-optical wavelength converters and signal regenerators", Proc. SPIE Asia-Pacific Optical Communications (APOC'05), Shanghai, China, paper 6021-95; Nov. 2005.

18. J. Leuthold, W. Freude, G. Boettger, **J. Wang**, P. Vorreau, A. Marculescu, "Trends in the field of all-optical wavelength conversion and regeneration for communication up to 160 Gb/s", Proc. European Conference on Optical Communications (ECOC'05), Glasgow, Scottland, Tu3.3.6; Sept. 2005 (invited)

19. A. Maitra, **J. Wang**, O. Hügel, C. G. Polton, W.Freude, J. Leuthold, "All-optical flip-flop based on an active stopband tapered DFB structure", Optical Amplifiers and Their Applications (OAA), WD3, Budapest, Aug. 2005.

20. W. Freude, C. G. Poulton, J. Brosi, Ch. Koos, F. Glöckler, **J. Wang**, M. Fujii, "Integrated optical waveguide devices: Design, modeling and fabrication", Proc. 16th Asia Pacific Microwave Conference (APMC'04), New Delhi, India; Dec. 2004 (invited).

21. C. G. Poulton, Ch. Koos, M. Müller, F. Glöckler, **J. Wang**, M. Fujii, J. Leuthold, W. Freude, "Sidewall roughness and deformations in high index-contrast waveguides and photonic crystals", Proc. 17th Annual Meeting of the IEEE Lasers and Electro-Optics Society (LEOS'04), Puerto Rico, USA; Nov. 2004.

22. W. Freude, J. Brosi, F. Glöckler, Ch. Koos, C. G. Poulton, **J. Wang**, M. Fujii, "High-index optical waveguiding structures", Proc. Optical Society of America Annual Meeting (OSA'04), paper TuS1, Rochester (NY), USA,; Oct. 2004 (invited).

23. W. Freude, C. G. Poulton, Ch. Koos, J. Brosi, F. Glöckler, **J. Wang**, G.-A. Chakam, M. Fujii, "Design and fabrication of nanophotonic devices", Proc. 6th Intern. Conf. on Transparent Optical Networks (ICTON'04), Mo.A.4 vol. 1, pp. 4-9, Wroclaw, Poland, July 2004 (invited).

Seminars and Workshops

1. W. Freude, M. Fujii, Ch. Koos, J.-M. Brosi, C. G. Poulton, **J. Wang**, J. Leuthold, "Wavelet FDTD methods and applications in nano-photonics", Proc. 386th WE Heraeus Seminar 'Computational Nano-Photonics' 2007, Bad Honnef, Feb. 25-28, 2007 (invited).

2. **J. Wang**, A. Maitra, W. Freude, J. Leuthold, "100 Gbit/s All-Optical Wavelength Conversion with an SOA and a Delay-Interferometer", Workshop der ITG Fachgruppe 5.3.1; "Modellierung photonischer Komponenten und Systeme", Technische Universität München, Germany Feb. 12-13, 2007.

3. J. Leuthold, W. Freude, **J. Wang**, P. Vorreau, A. Marculescu, "A novel all-optical DPSK wavelength converter based on Semiconductor Optical Amplifier (SOA) nonlinearities", Workshop der ITG Fachgruppe 5.3.1; "Modellierung photonischer Komponenten und Systeme " Lucent Technologies, Nürnberg, Germany, July 17-18, 2006.

4. J. Leuthold, W. Freude, G. Boettger, **J. Wang**, P. Vorreau, A. Marculescu, "Trends in the Field of All-Optical Wavelength Conversion and Regeneration", Workshop on Optical Transmission and Equalization (WOTE'05), Shanghai Jiao Tong University, Shanghai, China, Nov. 11, 2005.

5. A. Maitra, **J. Wang**, O. Hügel, C. G. Poulton, J. Leuthold, W. Freude, "Isolators and switches based on a nonlinear Bragg grating", Proc. European Cooperation in the Field of Scientific and Technical Research (COST'05), COST P11 Workshop and WG Meeting, Twente, Netherlands, Oct. 2-4, 2005.

6. C. G. Poulton, J. Brosi, Ch. Koos, F. Glöckler, **J. Wang**, M. Fujii, W. Freude, "Simulation, design and fabrication of integrated optical devices", Proc. Third Joint Symposium on Opto- and Microelectronic Devices and Circuits (SODC'04), Wuhan, China, March 2004.

7. W. Freude, J.-M. Brosi, G.-A. Chakam, F. Glöckler, Ch. Koos, C. G. Poulton, **J. Wang**, "Modelling and design of nanophotonic devices", Proc. URSI-Kleinheubacher Tagung, Miltenberg, Germany, Sep./Oct. 2003 (invited).

8. C. G. Poulton, J. Brosi, F. Glöckler, Ch. Koos, **J. Wang**, W. Freude, K. Il'in, M. Siegel, M. Fujii, "Modelling, design and realisation of SOI-based nanophotonic devices", Proc. Conference on Functional Nanostructures 2003 (CFN'03), Karlsruhe, Germany, Sep./Oct. 2003.

Supervision of Master Thesis (Diplomarbeit) and Study Project (Studienarbeit)

1. Zhenhao Zhang, *Pattern effect mitigation using red-shifted optical filter in semiconductor optical amplifier-based wavelength conversion*, study project, 2008.

2. Jingshi Li, *All-optical wavelength conversion based on a single semiconductor optical amplifier assisted by a pulse reformatting optical filter*, master thesis, 2007

3. Yang Jiao, *Analysis of optimal filtering technique for SOA based wavelength conversion*, master thesis, 2007.

4. Li Li, *Description of generalized non-linear loop mirror*, study project, 2007.

5. Muhammad Hafeez Chaudhary, *Study and simulation of optical 8-ary phase shift keying transmission system*, study project, 2006.

6. Jorge Hueso, *Characterization and optimization of a loop transmission experiment with a simulation tool*, master thesis, 2006.

7. Qiang Wang, *Embedding a numerical SOA model into an optical system simulator*, study project, 2005.

8. Yang Jiao, *Simulation and analysis of all-optical wavelength conversion of DPSK signals*, study project, 2005.

9. Oliver Hügel, *Modelling of all-optical signal processing in distributed feedback semiconductor optical amplifiers*, Diplomarbeit, 2004.

Acknowledgements

This work was partially funded by the research training group 786 "Mixed Fields and Nonlinear Interactions" of the Deutsche Forschungsgemeinschaft (German Research Foundation), in University of Karlsruhe (TH). Part of this work was supported by the Center for Functional Nanostructures (CFN) of the Deutsche Forschungsgemeinschaft within projects A3.1/A4.4, and by the European project TRIUMPH, Contract no. 027638.

Next, I would like to thank all those persons, who contributed directly or indirectly to the success of this work:

First of all, I would like to thank my supervisors, Prof. Dr. Dr. h. c. Wolfgang Freude and Prof. Dr. sc. nat. Jürg Leuthold of the Institute of High-Frequency and Quantum Electronics (IHQ) University of Karlsruhe, for granting me the opportunity to work on the topic of this dissertation. During my work at IHQ, Prof. Freude and Prof. Leuthold supported me as much as possible and have always been open for discussions and problems, both during the establishment of the numerical model and during the writing of the thesis.

I would like to thank Ayan Maitra and Andrej Marculescu for their many suggestions and constant support during this dissertation, especially during the early phases of chaos and confusion. I would especially thank Jingshi Li and Philipp Vorreau for many productive hours in the lab and inspiring discussions. I would also like to thank all my other colleagues at IHQ for their great support, René Bonk, Dr. Gunnar Böttger, Jan-Michael Brosi, Dr. Christian Koos, Dr. Shunfeng Li, Dr. Christopher Poulton, Moritz Röger, Dr. Stelios Sygletos, and Thomas Vallaitis.

Further, I would like to thank the personnel staff and the technique staff at IHQ, especially, Mrs. Lehmann, Mrs. Goldmann, Mrs. Kober and Mrs. Schubart. They created a pleasant and effective working environment for this dissertation.

I very much appreciate the project collaboration, either study project or master thesis, with Jingshi Li, Yang Jiao, Zhenhao Zhang, Li Li, Chen Wang, Qiang Wang, Oliver Hügel, Muhammad Hafeez Chaudhary, Jorge Hueso.

Last, but not least I wish to thank my parents for their everlasting support and to thank my wife, Xiaoting Li for her patience and love.

Curriculum Vitae

Name	Jin Wang
Date of birth	1st March 1972, Jiangsu, China
Citizenship	Chinese

Education and Professional Experience

Sep. 1989 – Jul. 1993	Study of electrical engineering (subject: radio technology) at University of Zhejiang, Hangzhou, China. Degree: Bachelor in Engineering
Aug. 1993 – 2000	Software engineer and later project manager in Alcatel Shanghai Bell Ltd., China
Sep. 1996 – Dez. 1996	Research in Belgium-Bell Ltd., Belgium
Apr. 2001 – Aug. 2003	Study of Electrical and Communication Engineering at the University of Karlsruhe, Karlsruhe, Germany. Degree: Diploma Engineering. Average note: 1.5 (very good) Diploma thesis work at the Institute of High-Frequency and Quantum Electronics.
Feb. 2003 – Aug. 2003	Topic: *Efficient techniques for modelling devices in integrated optics: Interpolet-collocation time-domain (ICTD) methods for large scale electromagnetic field analysis.* Note: 1.0 (very good)
Oct. 2003 – Mai 2008	PhD in Electrical and Communication Engineering from University of Karlsruhe, Karlsruhe, Germany. Thesis title: *Pattern Effect Mitigation Techniques for All-Optical Wavelength Converters Based on Semiconductor Optical Amplifiers.*

Awards and Achievements

- Chinese government award for outstanding self-financed students abroad, 2007.

- Awarded doctoral research fellowship (from June 2002, for 2.5 years), by Deutsche Forschungsemeinschaft (DFG) Research Training Group 786: Mixed Fields and Nonlinear Interactions (GKMF), for pursuing doctoral study in University of Karlsruhe, Karlsruhe, Germany.

- Awarded "Zuschluß zu den Kosten" by "Gesellschaft zur Pflege wissenschaftlicher Kontakte im Hause Heinrich Hertz" ("Heinrich-Hertz-Club"), June 2002.